Robert Elschner

Parametric Amplification and Wavelength Conversion

Robert Elschner

Parametric Amplification and Wavelength Conversion

A theoretical analysis with focus on phase-modulated signals

Südwestdeutscher Verlag für Hochschulschriften

Impressum/Imprint (nur für Deutschland/only for Germany)
Bibliografische Information der Deutschen Nationalbibliothek: Die Deutsche Nationalbibliothek verzeichnet diese Publikation in der Deutschen Nationalbibliografie; detaillierte bibliografische Daten sind im Internet über http://dnb.d-nb.de abrufbar.
Alle in diesem Buch genannten Marken und Produktnamen unterliegen warenzeichen-, marken- oder patentrechtlichem Schutz bzw. sind Warenzeichen oder eingetragene Warenzeichen der jeweiligen Inhaber. Die Wiedergabe von Marken, Produktnamen, Gebrauchsnamen, Handelsnamen, Warenbezeichnungen u.s.w. in diesem Werk berechtigt auch ohne besondere Kennzeichnung nicht zu der Annahme, dass solche Namen im Sinne der Warenzeichen- und Markenschutzgesetzgebung als frei zu betrachten wären und daher von jedermann benutzt werden dürften.

Verlag: Südwestdeutscher Verlag für Hochschulschriften GmbH & Co. KG
Heinrich-Böcking-Str. 6-8, 66121 Saarbrücken, Deutschland
Telefon +49 681 37 20 271-1, Telefax +49 681 37 20 271-0
Email: info@svh-verlag.de

Approved by: Berlin, TU, Diss., 2011

Herstellung in Deutschland:
Schaltungsdienst Lange o.H.G., Berlin
Books on Demand GmbH, Norderstedt
Reha GmbH, Saarbrücken
Amazon Distribution GmbH, Leipzig
ISBN: 978-3-8381-3076-7

Imprint (only for USA, GB)
Bibliographic information published by the Deutsche Nationalbibliothek: The Deutsche Nationalbibliothek lists this publication in the Deutsche Nationalbibliografie; detailed bibliographic data are available in the Internet at http://dnb.d-nb.de.
Any brand names and product names mentioned in this book are subject to trademark, brand or patent protection and are trademarks or registered trademarks of their respective holders. The use of brand names, product names, common names, trade names, product descriptions etc. even without a particular marking in this works is in no way to be construed to mean that such names may be regarded as unrestricted in respect of trademark and brand protection legislation and could thus be used by anyone.

Publisher: Südwestdeutscher Verlag für Hochschulschriften GmbH & Co. KG
Heinrich-Böcking-Str. 6-8, 66121 Saarbrücken, Germany
Phone +49 681 37 20 271-1, Fax +49 681 37 20 271-0
Email: info@svh-verlag.de

Printed in the U.S.A.
Printed in the U.K. by (see last page)
ISBN: 978-3-8381-3076-7

Copyright © 2012 by the author and Südwestdeutscher Verlag für Hochschulschriften GmbH & Co. KG and licensors
All rights reserved. Saarbrücken 2012

Für Katharina und Kasimir.

Contents

1 Introduction   5
   1.1 Future Challenges for Optical Networks 6
   1.2 Technologies for Parametric Amplification and Wavelength Conversion 9
   1.3 Goals of the Thesis 13

2 Modeling of Devices with Third-Order Nonlinearity   17
   2.1 Third-Order Nonlinear Materials 17
     2.1.1 The Third-Order Nonlinear Polarization 17
     2.1.2 Pulse Propagation in Nonlinear Media 19
     2.1.3 Third-Order Nonlinear Effects 22
   2.2 Highly Nonlinear Fibers 27
     2.2.1 Structure 27
     2.2.2 Pulse Propagation Equation 28
     2.2.3 Numerical Method 31
     2.2.4 Scattering Processes 32
     2.2.5 Nonideal Fiber Structure 35
   2.3 Semiconductor Optical Amplifiers (SOA) 36
     2.3.1 Structure 36
     2.3.2 Pulse Propagation Equation 37
     2.3.3 Gain Modeling 39
     2.3.4 Time-Domain Modeling 42

3 Phase-Modulation Formats   47
   3.1 Constellations 47
     3.1.1 Ideal Constellations 47
     3.1.2 Constellations in Presence of Noise 48
   3.2 Reception 49
     3.2.1 Direct Reception 49
     3.2.2 Coherent Reception 51
   3.3 Bit-Error Rate Estimation 53

	3.3.1	Additive White Gaussian Noise	53
	3.3.2	Deterministic Phase Distortions	55
	3.3.3	Nonlinear Phase Noise	56

4 Parametric Amplifiers and Wavelength Converters based on Four-Wave Mixing in HNLF — 57

- 4.1 General Characteristics 57
  - 4.1.1 Setup 58
  - 4.1.2 Conversion Efficiency and Conversion Spectrum 61
  - 4.1.3 Noise Figure 69
  - 4.1.4 Suppression of SBS by Pump Phase Modulation 71
  - 4.1.5 Additional Phase Distortions 74
- 4.2 Laser Phase Noise 78
  - 4.2.1 Single-Pump Configuration 79
  - 4.2.2 Dual-Pump Configuration 80
- 4.3 Impact of the Pump-Phase Modulation 81
  - 4.3.1 Single-Pump Configuration with Direct Detection 81
  - 4.3.2 Optical Compensation Using the Dual-Pump Conf. 85
  - 4.3.3 Optical Compensation Using the Single-Pump Conf. . . . 92
  - 4.3.4 Single-Pump and Dual-Pump Configurations with Coherent Detection 93
  - 4.3.5 Compensation Using Electronic Signal Processing 96
  - 4.3.6 Comparison to Impact on Amplitude Modulated Signals . 103
- 4.4 Pump-Induced Noise 104
  - 4.4.1 Pump-Induced Phase Noise in the Single-Pump Configuration 104
  - 4.4.2 Pump-Induced Phase Noise in the Dual-Pump Configuration 109
  - 4.4.3 Pump-Induced Amplitude Noise in the Single-Pump Configuration 112
  - 4.4.4 Pump-Induced Amplitude Noise in the Dual-Pump Configuration 115
- 4.5 Signal-Induced Phase Noise 117
  - 4.5.1 Single-Pump Configuration 119
  - 4.5.2 Dual-Pump Configuration 120

5 Wavelength Converters Based on Four-Wave Mixing in SOA — 123

- 5.1 General Characteristics 123
  - 5.1.1 Setup 123

	5.1.2 Conversion Efficiency	125
	5.1.3 Noise Figure	127
	5.1.4 Phase Distortions	132
5.2	Laser Phase Noise	132
5.3	Impact of Pump-Induced Noise	132
	5.3.1 Pump-Induced Phase Noise: Analytical Estimation	133
	5.3.2 Pump-Induced Phase Noise: Numerical Results	135
5.4	Impact of Signal-Induced Phase Noise	138
5.5	(O)SNR Penalty due to Pump- and Signal-Induced Phase Noise	141

6 Conclusions 147

A Definition of the Fourier Transforms 153

B Derivation of the Nonlinear Wave Equation 155

C Perturbation Theory 157

D Dispersion Characteristics 159

E Typical HNLF parameters 161

F Calculation of the FIR filter coefficients 163

G Simulation Parameters for SOA 165

H Analytical Solutions for FWM in HNLF 169
    H.1 Single Pump FWM 169
    H.2 Dual Pump FWM 171
        H.2.1 Phase Conjugation 171
        H.2.2 Frequency conversion 173

I BER Calculation for 16-QAM 175

J Phase Distortion after Carrier Phase Estimation 179

K Quadratic interpolation 183

L List of Acronyms 185

Chapter 1

Introduction

After the introduction of optical communication links in the late 1970s, their capacity has grown exponentially enabling today unprecedented possibilities for global communication via the Internet. To meet the continuing demand for growing capacity, optical networking will have to master two major challenges in the near future. Firstly, there is now a growing realization that the capacity will shortly reach the maximum limit which has been predicted theoretically for the current transmission link architecture shown in Fig 1.1 [1]. Secondly, the total power consumption of the Internet reaches today already one percent of the global power supply [2, 3]. The power consumption is dominated by switching and routing and will increase with increasing traffic. So, the possibility emerges that the Internet growth may ultimately be constrained by energy consumption rather than by capacity [4]. Mastering these two challenges requires the development of new technologies that help to push the maximum capacity limits while increasing the energy efficiency of the optical networks. In this chapter, it is discussed how parametric amplification and wavelength conversion can contribute to overcome the mentioned issues and which of the current technologies are the most promising to meet the requirements imposed by the system architecture. Finally, the goals of this thesis are summarized.

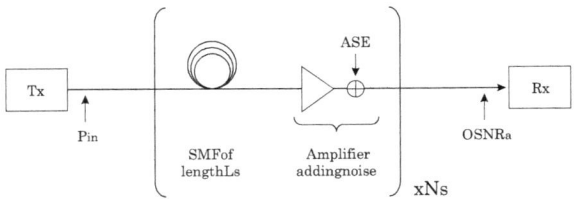

Figure 1.1: Simple transmission link architecture

1.1 Future Challenges for Optical Networks

There are several strategies to increase the capacity of the optical core network as the major building block of the Internet backbone. The most obvious method is to use more fibers. However, this option neither leads to increased cost- nor to increased energy efficiency, additionally complicates the network management and comprises high investment costs. Thus, a better strategy is to increase the capacity of the fiber itself. This can be done by either increasing the spectral efficiency of the used bandwidth, the increase of the used bandwidth or a better network efficiency. These options are discussed in the following and possible contributions of parametric amplifiers and wavelength converters are identified.

Increase of spectral efficiency

The fundamental limit for optical fiber transmission over links such as shown in Fig. 1.1 is given by the accumulation of amplified spontaneous emission (ASE) noise in the optical amplifiers. The signal quality after transmission is traditionally described by the optical signal-to-noise ratio (OSNR) [5, p. 65]

$$\text{OSNR} = \frac{P_{av}}{2P_{\text{ASE}}} \quad (1.1)$$

where P_{av} and P_{ASE} denote the average signal power and the ASE noise power in one polarization measured in the optical reference bandwidth (typically chosen as 12.5 GHz or equivalently 0.1 nm), respectively. The tolerance of the transmitted signal against ASE noise can be described by the required OSNR (labeled by OSNR_r) which denotes the minimum OSNR present after transmission in order to achieve a specified maximum bit error ratio (BER). The required OSNR depends on the used modulation format and the used symbol rate R_s. Knowing the required OSNR, the system design needs to ensure that the available OSNR at the receiver (labeled by OSNR_a) is sufficiently high. For a regular link design shown in Fig. 1.1, the available OSNR can be expressed in dB as [5, p. 67]

$$\text{OSNR}_a = 58 + P_{\text{in}} - \text{NF}_{\text{eff}} - L_s - 10\log_{10}(N_s) \quad (1.2)$$

58 dB is the OSNR for a quantum-noise limited signal with a power of 0 dBm. P_{in}, NF_{eff}, L_s and N_s denote the signal launch power (in dBm), the effective noise figure of the span (determined by the amplification scheme, in dB), the span loss (in dB) and the number of transmission spans, respectively.

Furthermore, the transmitted signal can be also degraded by other mechanisms like filtering, nonlinear effects etc. during transmission. In this case, the actual OSNR that is required at the receiver to achieve the specified BER can exceed $OSNR_r$ which only reflects the ASE noise tolerance. The difference in dB is called the OSNR penalty (labeled by $OSNR_{pen}$) and depends on the type of degradation.

Thus, the requirement for a successful transmission at a specified BER is

$$OSNR_a > OSNR_r + OSNR_{pen}. \qquad (1.3)$$

Increasing the spectral efficiency given in [bit/s/Hz] is achieved by using multi-level modulation formats like m-(differential) phase-shift keying (m-(D)PSK) or quadrature amplitude modulation (QAM) that encode several bits in one symbol. However, this lowers the ASE noise tolerance and thus increases the required OSNR [6]. To keep the same transmission distance, either the fiber loss or the effective noise figure of the span must be decreased or the launched power must be increased as shown in Eq. 1.2. Since the latter is accompanied by an increase of the OSNR penalty due to the onset of nonlinear effects, this is equivalent to the requirement to reduce either the fiber nonlinear coefficient or to compensate for nonlinear distortions.

The fiber loss of a standard single-mode fiber is mainly determined by material properties. Commercial systems operate in the C-band which is located around the minimum loss wavelength 1550 nm. To reduce the fiber loss significantly, a migration to hollow-core photonic crystal fibers was proposed which additionally provide extremely low nonlinear coefficients [1]. Investigations on the ultimate loss limit of these fibers have shown that their minimum loss may be located at longer wavelengths around 2000 nm [7]. Because the commercial and mature transmitter and receiver technology is available only around 1550 nm, transparent parametric wavelength converters are an excellent option to close this gap because they are able to convert whole wavelength bands in a single device [8]. Furthermore, parametric amplifiers are also an interesting option to provide amplification in this wavelength range due to the absence of erbium-doped fiber amplifiers (EDFAs) or distributed Raman amplification [9].

The effective noise figure NF of the span is determined by the optical amplification scheme. Phase-insensitive parametric amplifiers may be a viable alternative in reducing NF in comparison to erbium-doped fiber amplifiers because they provide a low noise figure close to the 3 dB quantum limit [10, 11]. Even more interesting (but also more challenging) are phase-sensitive parametric amplifiers that uniquely provide noise figures close to 0 dB [12, 13, 14].

Since parametric wavelength converters also provide optical phase conjugation (OPC), they can be used for compensation of the nonlinear effects using mid-span spectral inversion. Again, the possibility to convert wavebands makes this option attractive since only a single device per point-to-point transmission is necessary and intra- as well as inter-channel nonlinearities can be compensated [15, 16]. Furthermore, the advent of distributed Raman amplification offers now flat loss profiles along the fiber that are advantageous for the performance of the compensation of the nonlinearities [17].

Finally, ASE noise and nonlinear distortions can be also compensated for by all-optical regeneration. Parametric amplifiers can be used as amplitude using limiting amplifications as well as phase regenerators when operated as phase-sensitive amplifiers [18, 19, 20, 21, 22].

Increase of used bandwidth

The standard single-mode fiber provides a low loss transmission window of several hundred nm around the minimum loss wavelength of 1550 nm. Because the availability of optical amplification by erbium-doped fiber amplifiers (EDFA) is limited to the 35 nm wide C-band centered at 1550 nm, commercial systems deploying many wavelength division multiplexed (WDM) channels are still restricted to this wavelength range. To increase the usable bandwidth and thus the information capacity, optical amplification outside the C-band is necessary. Beside fiber amplifiers using other dopants than erbium and distributed Raman amplification, here fiber optic parametric amplifiers (FOPA) are an interesting option. They provide high gain optical amplification [23, 24] in bands with more than 50 nm bandwidth [25, 26, 27, 28] and, most importantly, arbitrary center wavelengths.

Increase of network efficiency

Packet switching, i.e. the method to group all transmitted data into suitably-sized blocks called packets which are then independently transmitted and switched, is a key component of the today's network structure. It provides an end-to-end connectivity and traffic grooming at the subwavelength level which presents a significant factor in ensuring a maximum utilization of network resources [29]. Up to now, electronic routers provide this functionality. However, with increasing channel data rates due to increasing symbol rates as well as the advent of multi-level modulation formats, transparent all-optical solutions like optical packet switching or optical burst switching (which uses much larger packets called bursts) are getting more attractive because of the limited speed of electronics [30]. In these scenarios, parametric wavelength converters provide key functions.

One important issue of all-optically packet switched networks is the lack of adequate optical buffering technology. Thus, optical burst switching was proposed. In this switching scheme, possible burst contentions are solved by wavelength conversion or by tunable delays. Both functions can be provided by transparent parametric wavelength converters [31].

One of the most promising options to realize a transparent optical switch fabric is based on arrayed waveguides (AWG). Because the AWG routes the signals in dependence on their wavelength, tunable wavelength converters are needed in front of the AWG. Due to their transparency and the possibility for waveband conversion, parametric wavelength converters (with option for integration) are also here the first choice. This switch fabric not only finds applications in all-optical networks, but also in hybrid opto-electronic packet routers that have been proposed to reduce the power consumption at the network nodes of the today's network architecture [2].

Performance monitoring is another issue in every transparent network, i.e. not only in all-optical networks but also in conventional networks providing optical bypasses at the core routers to reduce their power consumption [3]. Because parametric amplifiers inherently provide a wavelength converted copy of the amplified signal, they are ideal monitoring devices [32].

Finally, also ultra-fast all-optical digital logic is provided by parametric amplifiers and wavelength converters [33] which is a prerequisite for the realization of all-optical packet switched networks.

1.2 Technologies for Parametric Amplification and Wavelength Conversion

As shown in the previous section, parametric amplification and wavelength conversion can find a broad field of applications in future optical networks. A natural question is which requirements these devices have to fulfill to be actually useful and which technologies provide the necessary features. Table 1.1 shows a list of rather general requirements and useful features for amplifying and wavelength converting devices as well as more concrete target specifications as they can be estimated from today's perspective. Of course, it is impossible to fulfill all specifications with a single device, so that, depending on the application, just a subgroup of the listed points will be important. Because optical signal processing is a very active field of research, it would

Table 1.1: Requirements for and features of amplifying and wavelength converting signal-processing devices

Requirement/Feature	Target specification
Transparency to variable bit rates	1 Tbit/s
Transparency to amplitude and phase modulation formats	16 - QAM
Ability for waveband conversion	-
Ability for phase conjugation	-
Wide wavelength tunability	> C-band
Polarization independency	-
High gain/conversion efficiency	30 dB
Wideband flat gain /conversion spectrum	> C-band
Low noise figure	3 dB[1] / < 3 dB[2]
No amplitude or phase distortions	-
Low power consumption	-
Low coupling loss to transmission fiber	-
Suitable for photonic integration	-

[1] Phase-insensitive devices
[2] Phase-sensitive devices

Table 1.2: Classification of wavelength converter concepts after [34]

Concept	Nonlinear Effects	Bit rate transparency	Modulation format transparency
Opto-electronic	-	No	No[1]
Optical gating	XGM, XPM, saturable absorption, nonlinear loop mirror	Yes	No[2]
Wave mixing	DFG	Yes	Yes
	FWM	Yes	Yes
	SPM	Yes	No[3]
	electro-optical effect	No[4]	Yes
	acousto-optical effect	No[5]	Yes

[1] A coherent transceiver consisting of a coherent receiver and an IQ-modulator can be used in principle as a modulation-format transparent opto-electronic wavelength converter, provided that appropriate and flexible electronic circuitry (e.g. software-defined) is used.

[2] Recently, a modulation-format transparent wavelength converter based on XGM or XPM was proposed [35, 36].

[3] Not suitable for phase modulated signals due to generation of large excess phase noise [37]

[4] Because of small conversion bandwidth

[5] Because of very small conversion bandwidth

Table 1.3: Comparison of different wave mixing media

Process	Medium	Parametric amplification of CW signals shown	Suitable for photonic integration
FWM	Silica / Highly nonlinear fiber (HNLF)	Yes	No
	Soft glasses	No	No
	Indium phosphide (InP) / Semiconductor Optical Amplifier (SOA)	No	Yes (together with pump laser)
	Silicon waveguides	No	Yes
DFG	Periodically poled lithium niobate (PPLN)	No[1]	Yes

[1] Very recently demonstrated [38, 39]

be a desperate task to discuss the optimal technologies and devices for each application in detail. Thus, only some guidelines can be given. Among the requirements, the transparency regarding bit rate and modulation frequency are certainly at a very high priority in order to guarantee flexible use of the signal-processing device. Table 1.2 shows the classification of wavelength converting technologies after [34]. Among the different types, only wave-mixing (i.e. parametric) wavelength converters based on four-wave mixing (FWM) and difference-frequency generation (DFG) provide modulation format as well as bit rate transparency explaining the particular interest in these concepts. As DFG and FWM are nonlinear effects that originate from the $\chi^{(2)}$ and $\chi^{(3)}$ nonlinearity, respectively, they both occur in various media of which Table 1.3 shows a non-exhaustive list. The materials differ significantly in terms of e.g. linear and nonlinear loss and amount of nonlinearity leading to different device performance. Due to this, in particular the efficiency of the nonlinear effects differs significantly so that silica-based highly nonlinear fibers are the only [1] devices that offer continuous-wave parametric amplification [40, 41].

---

[1]Very recently, continuous-wave parametric amplification was also demonstrated in PPLN [38, 39].

Together with ultra-low splicing losses to a standard single-mode fiber, this fact also leads to a unique noise performance, as will be seen later on. With look on applications with many signal-processing devices, the suitability for integration is another natural distinctive criterion that plays a crucial role for the practicability and the commercialization perspective.

As was made plausible by the short overview, FWM in HNLFs and in SOAs were identified at the beginning of this thesis as two of the most promising devices for parametric amplification and wavelength conversion. The HNLF offers the best performance in terms of efficiency and noise performance while the SOA offers photonic integration, in particular with the pump lasers, which presents an advantage over the silicon waveguide and the PPLN. However, as the work presented within this thesis mainly comprises analytical findings, it can be of course generalized to other media. This will be used in the last chapter to come back to the list given in table 1.2 and draw some general conclusions.

1.3 Goals of the Thesis

Despite that modulation format transparency is a key feature for parametric wavelength converters in order to find applications in future optical networks, only a few investigations deal with phase-modulated signals or the question whether the conversion of amplitude and phase modulated signals is actually possible with the same device. The investigations of these issues was the main goal of this thesis. Particular attention has been paid to the identification of phase distortions and to the estimate of their impact in terms of BER on different higher-order phase modulation signals and QAM signals. Two of the most promising conversion media have been chosen. The HNLF provides the highest conversion efficiencies/ gain among all passive waveguides and easy coupling to the SSMF. On the other hand, the SOA as an active waveguide provides low size, low power consumption and the possibility for integration together with the pump laser.

Original contributions within this thesis include:

- The analytical derivation of the idler phase distortions due to the pump-phase modulation in single-stage and cascaded single-pump FOPAs and the semi-analytical calculation of the resulting OSNR penalties for direct detection formats [42].

- The analytical estimation of the tolerances for co- and counterphased pump-phase modulation used in dual-pump FOPAs to suppress the idler phase distortions and the semi-analytical calculation of the connected OSNR penalties for direct detection formats [43].

- The analytical derivation of the impact of the pump-phase modulation induced idler phase distortions on the coherent reception of m-PSK signals and the semi-analytical calculation of the connected OSNR penalties [44].

- The proposal and implementation of an algorithm to compensate for pump-phase modulation induced idler phase distortions in a coherent receiver and the characterization of its performance [45].

- The analytical derivation of the nonlinear phase noise variance due to XPM from a noisy pump in FOPAs and the semi-analytical calculation of the related OSNR penalty for direct and coherent detection formats [46].

- The analytical derivation of the variance and statistics of nonlinear amplitude noise generated by gain fluctuations due to noisy pumps in FOPAs and the semi-analytical calculation of the related OSNR penalty for 16-QAM signals [47].

- The numerical estimation of the nonlinear phase noise variance due to a noisy pump in SOA-based wavelength converters and the semi-analytical calculation of the related OSNR penalty for direct and coherent detection formats [48].

- The numerical characterization of the nonlinear noise transfer from the pump to the signal and the idler as well as from the signal to the idler in a SOA-based wavelength converter.

Although this thesis includes only theoretical results, it is important to mention that virtually all results have been also experimentally confirmed by project partners. Citations on their work will be given in the individual sections.

The thesis is structured as follows: Chapter 2 covers the modeling of $\chi^{(3)}$-media. The general pulse propagation equation is derived and the nonlinear effects are discussed. Then, the particular pulse propagation equations for the HNLF and the SOA are derived and its numerical evaluation is explained. Chapter 3 discusses the constellations, the reception and the BER estimation for higher-order phase modulation and QAM formats. In Chapter 4, parametric amplifiers and wavelength converters based on HNLF are treated. The general characteristics are discussed and the different phase distortions identified. Then the impact of the phase distortions on the BER is estimated for different modulation formats. Similarly, Chapter 5 treats parametric wavelength converter based on SOAs. Finally, Chapter 6 provides a summary and the conclusions.

Chapter 2

Modeling of Devices with Third-Order Nonlinearity

Nonlinear optics is the study of phenomena that occur as a consequence of the modification of the optical properties of a material system by the presence of light [49, p. 1]. Thus, for a correct description of nonlinear devices, not only the light propagation governed by the Maxwell equations, but also the light-matter interaction has to be accounted for, typically using phenomenological models to keep the model complexity low. In this chapter, the nonlinear wave equation is introduced and the different nonlinear effects related to the third-order nonlinearity are discussed. Furthermore, the highly nonlinear fiber (HNLF) and the semiconductor optical amplifier (SOA) are discussed as nonlinear devices and their phenomenological models are presented in detail which will be used in the later chapters for the simulation of the HNLF- and SOA-based parametric amplifiers and wavelength converters.

2.1 Third-Order Nonlinear Materials

2.1.1 The Third-Order Nonlinear Polarization

The propagation of light in a nonlinear medium is governed by the nonlinear wave equation [see Appendix B for the full derivation],

$$\left(\Delta - \frac{1}{c_0^2}\frac{\partial^2}{\partial t^2}\right)\vec{E} = \frac{1}{\epsilon_0 c_0^2}\frac{\partial^2 \vec{P}}{\partial t^2} \tag{2.1}$$

Here, $\vec{E}$ denotes the vectorial electric field, c_0 is the velocity of light in vacuum, ϵ_0 is the vacuum permittivity and $\vec{P}$ is the polarization of the medium. To complete the description, a relation between $\vec{P}$ and $\vec{E}$ is needed. In general, this requires a quantum-mechanical approach to account for the atomistic structure of the medium [50, p. 26]. In practice, one often uses phenomenological models due to their reduced complexity. For SOAs and silica fibers, these models differ substantially because the light interacts resonantly and non-resonantly with the media in the wavelength range of interest, respectively. The consequences for the polarization will be briefly discussed in the following.

Non-resonant Nonlinearities

In silica fibers, the light mainly interacts with bound electrons in the wavelength range of interest, i.e. the interaction is non-resonant [50, p. 26]. In this case, the nonlinearity is weak and the polarization can be expanded into a quickly converging power series that can be written symbolically as

$$\vec{P} = \epsilon_0 \left(\chi^{(1)} : \vec{E} + \chi^{(2)} : \vec{E}^2 + \chi^{(3)} : \vec{E}^3 + ... \right) \tag{2.2}$$

where the operators $\chi^{(n)}$ are called nth-order susceptibilities. This relation can be simplified using some material properties of silica: First, silica is an isotropic material that possesses inversion symmetry. Second, the linear medium response is local and frequency-dependent and the nonlinear medium response is local and frequency-independent, i.e. instantaneous. This assumption is justified since the nonlinear response time from the bound electrons is in the order of 10^{-15} s. Additionally, the electric field shall be linearly polarized in x-direction. Then, the vectorial relation Eq. (2.2) reduces to the scalar relation [49, p. 38, p. 44, p. 53, p. 56]

$$P(\vec{r},t) = \underbrace{\epsilon_0 \int_{-\infty}^{\infty} \chi_{xx}^{(1)}(t-t') E(\vec{r},t') dt'}_{P^{(1)}} + \underbrace{\epsilon_0 \chi_{xxxx}^{(3)} E(\vec{r},t)^3}_{P^{(3)}} + ... \tag{2.3}$$

with the linear polarization $P^{(1)}$ and the lowest-order (third-order) nonlinear polarization $P^{(3)}$. If one considers a monochromatic field,

$$E(\vec{r},t) = \frac{1}{2} \hat{E}(\vec{r}) e^{-i\omega_0 t} + \text{c.c.}, \tag{2.4}$$

the nonlinear polarization takes the form

$$P^{(3)} = \frac{1}{8} \epsilon_0 \chi_{xxxx}^{(3)} \hat{E}(\vec{r})^3 e^{-3i\omega_0 t} + \frac{3}{8} \epsilon_0 \chi_{xxxx}^{(3)} |\hat{E}(\vec{r})|^2 \hat{E}(\vec{r}) e^{-i\omega_0 t} + \text{c.c.}. \tag{2.5}$$

Thus, the monochromatic field creates a nonlinear polarization oscillating at the two distinct frequency components $3\omega_0$ and ω_0. The first term leads to the process of third-harmonic generation and can be generally neglected in silica fibers [50, p. 33]. Then, the total polarization of the medium can be written as

$$P(\vec{r},t) = \epsilon_0 \int_{-\infty}^{\infty} \chi_{xx}^{(1)}(t-t') E(\vec{r},t') dt' + \frac{3}{4} \epsilon_0 \chi_{xxxx}^{(3)} |\hat{E}(\vec{r})|^2 E(\vec{r},t). \tag{2.6}$$

A medium which exhibits a nonlinear polarization of the form shown in Eq. (2.6) will be referred in the following as third-order nonlinear medium. If the nonlinear susceptibility is real-valued as it is the case for the HNLF, the medium is called a Kerr medium.

Resonant Nonlinearities

In SOAs, the light-medium interaction is dominated by photon-induced transitions of electrons between different energy bands, i.e. the interaction is resonant. In this case, the nonlinear polarization is strong and the power series expansion in Eq. (2.2) does not always converge [49, p. 277]. Furthermore, the response is relatively slow in the order of 10^{-10} s and therefore strongly frequency-dependent. If the monochromatic field from Eq. (2.4) is applied, the steady state resonant polarization is given by [49, p. 277]

$$P_r(\vec{r},t) = \frac{\epsilon_0 \chi_r(\omega_0)}{1+\left|\hat{E}(\vec{r})\right|^2/|E_s|^2} E(\vec{r},t), \qquad (2.7)$$

where $|E_s|^2$ is the saturation intensity. $\chi_r(\omega_0)$ is the complex linear susceptibility in the case of a weak field. Eq. (2.7) can be expanded into

$$P_r(\vec{r},t) = \epsilon_0 \chi_r(\omega_0) \left(1 - \frac{\left|\hat{E}(\vec{r})\right|^2}{|E_s|^2} + \left(\frac{\left|\hat{E}(\vec{r})\right|^2}{|E_s|^2}\right)^2 + \ldots \right) E(\vec{r},t), \qquad (2.8)$$

however, this series only converges if $\left|\hat{E}(\vec{r})\right|^2 < |E_s|^2$. In this limit, it is valid to truncate the power series after the second summand making Eq. (2.8) formally equivalent to Eq. (2.6). Then, the SOA behaves like a third-order nonlinear medium.

2.1.2 Pulse Propagation in Nonlinear Media

In the following, the propagation of linearly polarized pulses in a nonlinear waveguide medium will be treated. For that purpose, the scalar version of Eq. (2.1) is transformed into Fourier space,

$$\left(\Delta + \frac{\omega^2}{c_0^2}\right) \tilde{E}(\vec{r},\omega) = \frac{\omega^2}{\epsilon_0 c_0^2} \tilde{P}(\vec{r},\omega). \qquad (2.9)$$

The electric field shall be given by

$$E(\vec{r},t) = \frac{1}{2} c_0 \epsilon_0 \, C \, A(z,t) \, F(x,y) \, e^{i(\beta_0 z - \omega_0 t)} + c.c. \qquad (2.10)$$

with the slowly varying envelope $A(z,t)$ and the transverse profile $F(x,y)$. The normalization constant ensures that the optical power is given by $|A(z,t)|^2$, i.e.

$$C^2 = \left(\frac{1}{2} \int_{-\infty}^{\infty} \int_{-\infty}^{\infty} |F(x,y)|^2 \, dx dy \right)^{-1}. \qquad (2.11)$$

The Fourier transforms of $E(\vec{r},t)$ and $A(z,t)$ are connected by

$$\begin{aligned}
\tilde{E}(\vec{r},\omega) &= \frac{1}{2} C \, \tilde{A}(z, \omega - \omega_0) \, F(x,y) \, e^{i\beta_0 z} + \frac{1}{2} C \, \tilde{A}^*(z, \omega + \omega_0) \, F^*(x,y) \, e^{-i\beta_0 z} \\
&\cong \frac{1}{2} C \, \tilde{A}(z, \omega - \omega_0) \, F(x,y) \, e^{i\beta_0 z} \qquad (2.12)
\end{aligned}$$

The approximation can be made since a quantity that varies slowly in time cannot posses high frequency components. The Fourier transform of the polarization $\tilde{P}$ shall be given by

$$\tilde{P}(\vec{r},\omega) = \epsilon_0 \tilde{\chi}(x,y,\omega) \tilde{E}(\vec{r},\omega) \qquad (2.13)$$

where the frequency domain susceptibility $\tilde{\chi}(x,y,\omega)$ is dependent on x and y because of the spatially varying waveguide cross section. It shall include the linear as well as the nonlinear material response. Strictly speaking, the latter is generally not possible due to the form of Eq. (2.6) or Eq. (2.7). However, the approach is justified since the nonlinearities will be treated as a small perturbation as it is explained later [50, p. 33]. Insertion of Eqs. (2.12) and (2.13) in Eq. (2.9) yields

$$\left(\frac{\partial^2}{\partial x^2} + \frac{\partial^2}{\partial y^2} + 2i\beta_0 \frac{\partial}{\partial z} + \frac{\omega^2}{c_0^2} \epsilon(x,y,\omega) - \beta_0^2 \right) \tilde{A}(z, \omega - \omega_0) \, F(x,y) = 0 \qquad (2.14)$$

where $\frac{\partial^2}{\partial z^2} \ll \beta_0^2$ was used. Additionally, the complex dielectric constant $\epsilon(x,y,\omega) = 1 + \tilde{\chi}(x,y,\omega)$ was defined. Using the method of separation of the variables, Eq. (2.14) can be decomposed into two separate equations for A and F [50, p. 34],

$$\left(\frac{\partial^2}{\partial x^2} + \frac{\partial^2}{\partial y^2} + \frac{\omega^2}{c_0^2} \epsilon(x,y,\omega) - \tilde{\beta}^2 \right) F(x,y) = 0 \qquad (2.15)$$

$$\left(2i\beta_0 \frac{\partial}{\partial z} + \tilde{\beta}^2 - \beta_0^2 \right) \tilde{A}(z, \omega - \omega_0) = 0. \qquad (2.16)$$

The eigenvalue equation (2.15) is solved by first-order perturbation theory [50, p. 34], [51, p. 40]. The complex ϵ is split into $\epsilon(x,y,\omega) = \epsilon_b(x,y,\omega) + \Delta\epsilon(x,y,\omega)$, where $\epsilon_b(x,y,\omega)$ is the (real-valued) background dielectric constant due to the linear material response (that contains the spatially varying dielectric profile of the waveguide) and $\Delta\epsilon(x,y,\omega)$ is a (complex-valued) perturbation that shall include all imaginary and nonlinear parts. The idea is that the transverse profile $F(x,y)$ is determined by $\epsilon_b(x,y,\omega)$ while $\Delta\epsilon(x,y,\omega)$ acts as a small perturbation changing only the propagation constant which splits into $\tilde{\beta}(\omega) =$

$\beta(\omega) + \Delta\beta(\omega)$. $F(x,y)$ and $\beta(\omega)$ are obtained by solving Eq. (2.15) with $\epsilon(x,y,\omega) = \epsilon_b(x,y,\omega)$. $\Delta\beta(\omega)$ is given by (see Appendix C)

$$\Delta\beta(\omega) = \frac{\omega^2}{2\beta(\omega)c_0^2} \frac{\int_{-\infty}^{\infty}\int_{-\infty}^{\infty} \Delta\epsilon(x,y,\omega)|F(x,y)|^2\,dxdy}{\int_{-\infty}^{\infty}\int_{-\infty}^{\infty} |F(x,y)|^2\,dxdy}. \tag{2.17}$$

Now, the propagation constant can be inserted into Eq. (2.16) yielding

$$\left(i\frac{\partial}{\partial z} + \beta(\omega) + \Delta\beta(\omega) - \beta_0\right)\tilde{A}(z,\omega-\omega_0) = 0. \tag{2.18}$$

where $\tilde{\beta}^2 - \beta_0^2$ was approximated by $2\beta_0(\tilde{\beta}-\beta_0)$. In order to transform Eq. (2.18) back into the time domain, $\beta(\omega)$ is expanded into a Taylor series around ω_0,

$$\beta(\omega) = \sum_{n=0}^{4} \frac{\beta_n}{n!}(\omega-\omega_0)^n \tag{2.19}$$

with the coefficients

$$\beta_n = \left.\frac{d^n\beta}{d\omega^n}\right|_{\omega=\omega_0}. \tag{2.20}$$

The terms with order higher than 4 are neglected. The relationship of these dispersion coefficients with the experimentally measurable dispersion $\underline{D}$ is given in Appendix E. Similarly,

$$\Delta\beta(\omega) \approx \Delta\beta(\omega_0) \tag{2.21}$$

is expanded keeping only the zeroth order term. Then, the Fourier transform of Eq. (2.18) gives

$$\left(i\frac{\partial}{\partial z} + \sum_{n=2}^{4}\frac{\beta_n}{n!}(i)^n\frac{\partial^n}{\partial T^n} + \Delta\beta(\omega_0)\right)A(z,T) = 0. \tag{2.22}$$

In the last step, the transformation [50, p. 40]

$$T = t - z/v_G = t - \beta_1 z \tag{2.23}$$

was applied that defines a retarded time T within a reference frame moving with the group velocity v_G. Eq. 2.22 is the final result of this section and describes the propagation of pulses in a dispersive and nonlinear medium. the coefficients β_n describe the dispersion and $\Delta\beta(\omega_0)$ includes the absorption and the nonlinearities. In the following sections, these parameters will be specified first for a general third-order nonlinear medium and later for the HNLF and the SOA.

2.1.3 Third-Order Nonlinear Effects

In this section, Eq. (2.22) is used to further analyze the pulse propagation in a rather general third-order nonlinear material. The polarization is given by Eq. (2.6) and the complex dielectric constant is given by

$$\epsilon(x,y,\omega) = \underbrace{1 + \Re\{\tilde{\chi}_{xx}^{(1)}\}}_{\epsilon_b(x,y,\omega)} + \underbrace{i\Im\{\chi_{xx}^{(1)}\} + \frac{3}{4}\chi_{xxxx}^{(3)}|\hat{E}(x,y)|^2}_{\Delta\epsilon(x,y)}. \quad (2.24)$$

Insertion into Eq. 2.17 gives

$$\Delta\beta(\omega_0) = \frac{\omega_0^2}{2\beta_0 c_0^2} \frac{\int_{-\infty}^{\infty}\int_{-\infty}^{\infty} \Delta\epsilon(x,y)|F(x,y)|^2 dxdy}{\int_{-\infty}^{\infty}\int_{-\infty}^{\infty} |F(x,y)|^2 dxdy}. \quad (2.25)$$

On the other hand, it is convenient to express $\Delta\epsilon(x,y)$ as a function of the intensity dependent refractive index and absorption coefficient that can be defined as

$$\underline{n}(x,y,\omega) = n_0(\omega) + n_2|\hat{E}(x,y)|^2 \quad (2.26)$$

$$\alpha = \alpha_0 + \alpha_2|\hat{E}(x,y)|^2 \quad (2.27)$$

Thereby, n_2 and α_2 are the third-order nonlinear refractive index and the third-order nonlinear absorption, respectively. $n_0(\omega)$ is defined by the propagation constant, $\beta(\omega) = n_0(\omega)\omega/c_0$, which was obtained by solving Eq. (2.15). The square of the complex refractive index at ω_0 is given by

$$\left(\underline{n}(x,y,\omega_0) + i\frac{\alpha c_0}{2\omega_0}\right)^2 \cong \quad (2.28)$$

$$n_0(\omega_0)^2 + \underbrace{2n_0(\omega_0)n_2|\hat{E}(x,y)|^2 + i\frac{n_0(\omega_0)c_0}{\omega_0}(\alpha_0 + \alpha_2|\hat{E}(x,y)|^2)}_{\Delta\epsilon(x,y)}.$$

where $n_0 \gg n_2|\hat{E}|^2$ and $n_0 \gg \alpha c_0/(2\omega_0)$ was used. A comparison between Eqs. 2.24 and 2.28 yields

$$n_2 = \frac{3}{8n_0(\omega_0)}\Re\{\chi_{xxxx}^{(3)}\} \quad (2.29)$$

$$\alpha_0 = \frac{\omega_0}{c_0 n_0(\omega_0)}\Im\{\chi_{xx}^{(1)}\} \quad (2.30)$$

$$\alpha_2 = \frac{3\omega_0}{4c_0 n_0(\omega_0)}\Im\{\chi_{xxxx}^{(3)}\}. \quad (2.31)$$

Insertion into Eq. (2.25) gives

$$\Delta\beta(\omega_0) = i\frac{\alpha_0}{2} + \left(\omega_0 n_2/c_0 + i\frac{\alpha_2}{2}\right)\frac{|A(z,T)|^2}{A_{eff}}. \quad (2.32)$$

where the effective mode area is defined by [50, p. 35]

$$A_{eff} = \frac{\left(\int_{-\infty}^{\infty}\int_{-\infty}^{\infty} |F(x,y)|^2 \, dx\, dy\right)^2}{\int_{-\infty}^{\infty}\int_{-\infty}^{\infty} |F(x,y)|^4 \, dx\, dy}. \tag{2.33}$$

Thus, Eq. (2.22) takes the form

$$\left(i\frac{\partial}{\partial z} + \sum_{n=2}^{4} \frac{\beta_n}{n!}(i)^n \frac{\partial^n}{\partial T^n} + i\frac{\alpha_0}{2} + \tilde{\gamma}|A(z,T)|^2\right) A(z,T) = 0. \tag{2.34}$$

with the complex nonlinear coefficient $\tilde{\gamma} = (\omega_0 n_2/c_0 + i\frac{\alpha_2}{2})/A_{eff}$. Eq. 2.34 comprises two different types of nonlinearity. The index nonlinearity is related to the real part of $\tilde{\gamma}$ and impacts only the phase of the propagating wave. In contrast, the imaginary part of $\tilde{\gamma}$ is related to the gain nonlinearity that impacts the absorption or the amplification of the propagating wave. To illustrate all nonlinear effects that may occur the slowly varying envelope $A(z,T)$ shall consist of three distinct waves with different center frequencies,

$$A(z,T) = \sum_{l=1}^{3} A_l(z,T) \, e^{i(B_l z - \Omega_l t)}. \tag{2.35}$$

Thereby, $\Omega_l = \omega_l - \omega_0 \ll \omega_0$ are difference frequencies relative to the reference frequency and $B_l = \beta(\omega_l) - \beta_0$ are difference wavenumbers. Insertion in the nonlinear part of Eq. (2.34) leads to the following expression:

$$|A|^2 A = \sum_{l=1}^{3} |A_l|^2 A_l \, e^{i(B_l z - \Omega_l t)} \qquad \text{(SPM/SGM)}$$

$$+ 2 \sum_{\substack{l,m=1 \\ l \neq m}}^{3} |A_m|^2 A_l \, e^{i(B_l z - \Omega_l t)} \qquad \text{(XPM/XGM)}$$

$$+ \sum_{\substack{l,m=1 \\ l \neq m}}^{3} A_m^2 A_l^* \, e^{i[(2B_m - B_l)z - (2\Omega_m - \Omega_l)t]} \qquad \text{(DFWM)}$$

$$+ \sum_{\substack{l,m,n=1 \\ l \neq m \neq n}}^{3} A_m A_n A_l^* \, e^{i[(B_m + B_n - B_l)z - (\Omega_m + \Omega_n - \Omega_l)t]} \qquad \text{(NDFWM)}$$

The nonlinear interaction of the three waves generates 18 different terms. The first nine can be arranged into groups of 3 having the same frequencies as the incident waves. These terms cause self-phase/self-gain modulation (SPM/SGM) and cross-phase/cross-gain modulation (XPM/XGM) leading to nonlinear phase shifting and nonlinear absorption/gain. In the second nine terms, new frequency components are created. These terms cause degenerate four-wave mixing (DFWM) and nondegenerate four-wave mixing (NDFWM) and act as source terms in Eq. (2.34).

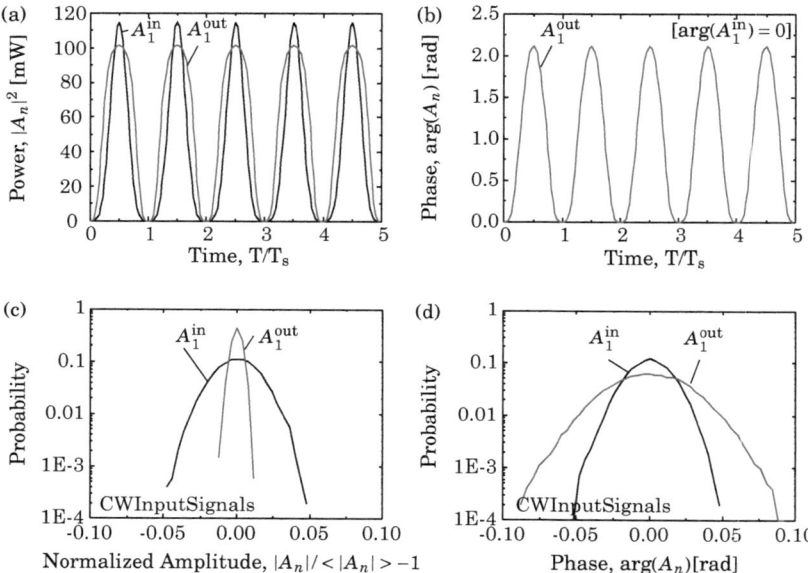

Figure 2.1: Effects of SGM and SPM on a 10 GHz RZ-33 pulse train A_1 with an average power of 16 dBm: (a) Power of the input and output field A_1, (b) Phase of the input and output field A_1. Effects of SGM and SPM on a noisy CW signal A_1 with an average power of 17 dBm and an OSNR of 30 dB: (c) Power histogram of the input and output field, (d) Phase histogram of the input and output field A_1. The calculations have been performed using Eq. 2.34 with the following parameters: L = 1 m, α_0 = -2/m, $\beta_2 = \beta_3 = \beta_4 = 0$, $\tilde{\gamma} = (-20 - i\,10)/(m\,W)$

Self-Gain Modulation (SGM)

SGM is caused by the gain nonlinearity and is a direct consequence of the intensity-dependent absorption/gain given in Eq. (2.27). It is also referred to as absorption/gain saturation. Fig. (2.1 (a)) shows how SGM flattens pulses due the lower gain at the pulse peaks.

Self-Phase Modulation (SPM)

SPM is caused by the index nonlinearity and is a direct consequence of the intensity-dependent refractive index given in Eq. (2.26). Fig. (2.1 (b)) shows how SPM modulates the phase of pulses. In the presence of amplitude noise, SPM leads to the generation of nonlinear phase noise as shown in Fig. (2.1 (d)).

Cross-Gain Modulation (XGM)

XGM has the same origin as SGM. If one distinguishes different incident waves, then XGM describes the gain or absorption change for every wave that occurs due to the presence of the other waves. Fig. (2.2 (a)) shows how the amplitude of a signal A_1 is modulated by a pulse train A_2. If A_2 is degraded by amplitude noise, XGM generates nonlinear amplitude noise at A_1 as shown in Fig. (2.2 (c)).

Cross-Phase Modulation (XPM)

XPM has the same origin as SPM. If one distinguishes different incident waves, then XPM describes the phase shift on every wave that occurs due to the presence of the other waves. Fig. (2.2 (b)) shows how the phase of a signal A_1 is modulated by a pulse train A_2. Similarly to SPM, also XPM leads to the generation of nonlinear phase noise in A_1 if A_2 is degraded by amplitude noise as shown in Fig. (2.2 (d)).

Four-Wave Mixing (FWM)

FWM is caused by the gain as well as by the index nonlinearity. It generates new frequency components with frequencies

$$\Omega_d = \Omega_a + \Omega_b - \Omega_c. \tag{2.36}$$

where the indices a, b, c are arbitrarily chosen from the input wave indices $\{1,2,3\}$ such that never $a = b = c$. These are shown in Fig. (2.3). If $a \neq b \neq c$, the process in called non-degenerate FWM (NDFWM) while for $a = b \neq c$ it is

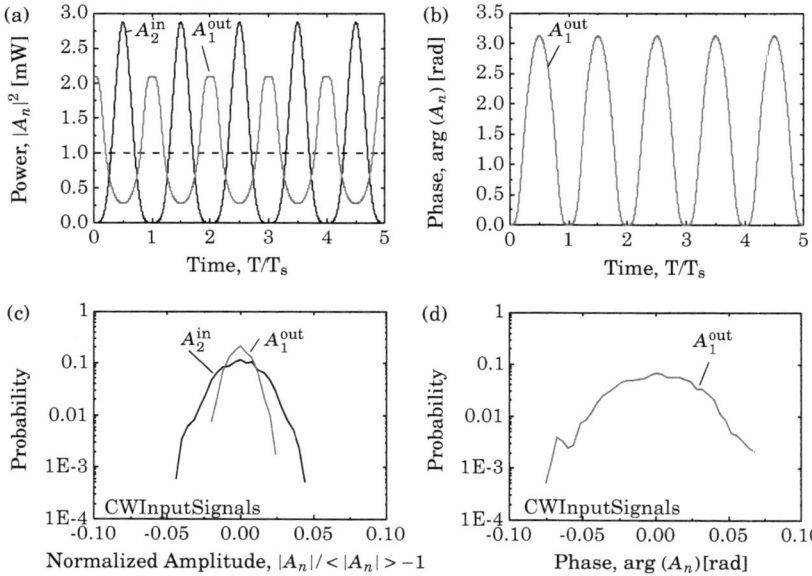

Figure 2.2: Effects of XGM and XPM on a noiseless CW signal A_1 with an average power of -10 dBm in presence of a 10 GHz RZ-33 pulse train A_2 with an average power of 16 dBm, the signals were separated by 500 GHz: (a) Power of the input field A_2 and output field A_1, (b) Phase of the input field A_2 and output field A_1. Effects of XGM and XPM on a noiseless CW signal A_1 with an average power of -10 dBm in presence of a noisy CW signal A_2 with an average power of 17 dBm and an OSNR of 30 dBm, the signals were separated by 500 GHz: (c) Power histogram of the input field A_2 and output field A_1, (d) Phase histogram of the input field A_1 and output field A_2. The calculations have been performed using Eq. 2.34 with the following parameters: L = 1 m, α_0 = -2/m, $\beta_2 = \beta_3 = \beta_4 = 0$, $\tilde{\gamma}$ = (-20 - i 10)/(m W).

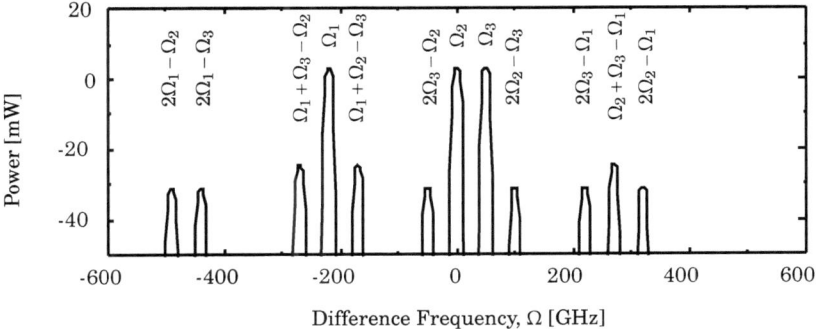

Figure 2.3: Generation of new frequency components due to FWM of three input CW waves with frequencies Ω_1, Ω_2 and Ω_3. The average power per signal was -5 dBm and the frequency separations were $\Omega_3 - \Omega_2 = 50$ GHz and $\Omega_2 - \Omega_1 = 220$ GHz. The calculations have been performed using Eq. 2.34 with the following parameters: L = 1 m, α_0 = -2/m, $\beta_2 = \beta_3 = \beta_4 = 0$, $\tilde{\gamma}$ = (-20 - i 10)/(m W)

called degenerate FWM (DFWM). The new frequency component is created by an energy transfer from the waves a and b to the waves c and d. This energy transfer is only efficient if

$$\Delta B = B_c + B_d - B_a - B_b \cong 0 \tag{2.37}$$

which is referred to as the phase-matching condition. ΔB is called the linear phase mismatch. The fulfillment of Eq. 2.37 typically requires low chromatic dispersion. If the efficiency is so high that wave c exhibits significant amplification the process is also called parametric amplification. In the photon picture, FWM can be understood as a nonlinear process in which two photons with energies $\hbar\Omega_a$ and $\hbar\Omega_b$ are annihilated and two other photons with energies $\hbar\Omega_c$ and $\hbar\Omega_d$ are created. Then, Eqs. (2.36) and (2.37) just represent the energy and impulse conservation.

2.2 Highly Nonlinear Fibers

2.2.1 Structure

The optical fiber is a circular waveguide made of fused silica (SiO_2) that guides light due to total internal reflexion. Its structure is shown schematically in Fig. (2.4a). The simplest form of the refractive index profile is shown in Fig. (2.4b). It is called the step index profile and consists of a core with a refractive

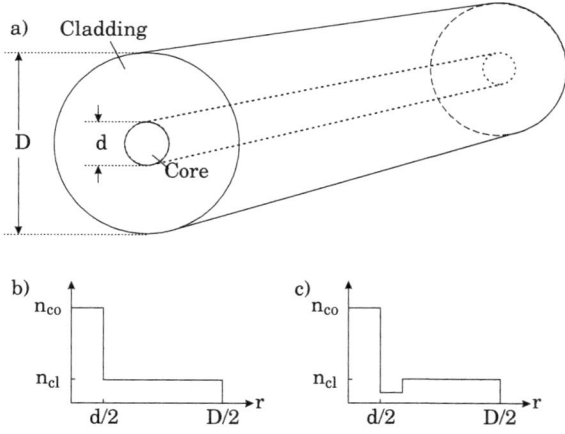

Figure 2.4: Structure of an optical fiber

index n_{co} and a cladding with n_{cl} where total internal reflexion requires

$$n_{co} > n_{cl}. \tag{2.38}$$

The step index profile has two degrees of freedom, i.e. the core radius $d/2$ and the relative index step, $\Delta n = (n_{co} - n_{cl})/n_{co} \ll 1$. It is up to some percent in the HNLF and is realized by doping the core with germanium dioxide (GeO_2) to increase the refractive index. State-of-the-art highly nonlinear fibers typically possess a more complicated index profile, the so-called W-shaped profile shown in Fig. (2.4c) [52, 40]. The inner cladding ring with the depressed refractive index due to doping with fluorine (F) gives two more degrees of freedom in fiber design which can be used to optimize dispersive and nonlinear properties of the fiber at the same time.

The fabrication of such fibers is done in two steps. First, a cylindrical preform with the desired index profile and the relative core-cladding dimensions is prepared using a chemical vapor-deposition method. Afterwards, the preform is drawn into a fiber by feeding it into a furnace with proper speed. During this process, the index profile and the relative core-cladding dimensions are maintained [50, p. 4].

2.2.2 Pulse Propagation Equation

As discussed in section 2.1.1, the HNLF is a Kerr medium, i.e. it is a third-order nonlinear medium with a real-valued nonlinear susceptibility. Thus, the pulse propagation is described by Eq. 2.34,

$$\left(i\frac{\partial}{\partial z} + \sum_{n=2}^{4}\frac{\beta_n}{n!}(i)^n\frac{\partial^n}{\partial T^n} + i\frac{\alpha_0}{2} + \gamma|A(z,T)|^2\right)A(z,T) = 0, \quad (2.39)$$

with the real-valued nonlinear coefficient

$$\gamma = \omega_0 n_2/(c_0 A_{eff}). \quad (2.40)$$

In the literature, this equation is also called (generalized) Nonlinear Schrödinger (NLS) equation [50, p. 40].

Chromatic Dispersion In order to calculate the chromatic dispersion and the nonlinear coefficient of the fiber, the propagation constant $\beta(\omega)$ and the transversal field profile $F(x,y)$ have to be determined. For this aim, Eq. 2.15 has to be solved using the appropriate circular refractive index profile. To generalize the solution, Eq. 2.15 is typically normalized to obtain the relation $\underline{B}(\underline{V})$ instead of $\beta(\omega)$ with the normalized frequency [53, p. 128], [54, p. 38]

$$\underline{V} = k_0\frac{d}{2}\sqrt{n_{co}^2 - n_{cl}^2} \cong k_0\frac{d}{2}n_{co}\sqrt{2\Delta n} \quad (2.41)$$

and the normalized propagation constant

$$\underline{B} = \frac{\frac{\beta^2}{k_0^2} - n_{cl}^2}{n_{co}^2 - n_{cl}^2} \approx \frac{\frac{\beta}{k_0} - n_{cl}}{n_{co} - n_{cl}}. \quad (2.42)$$

$k_0 = \omega/c_0$ is the vacuum propagation constant. Although Eq. 2.15 is usually solved numerically for a general refractive index profile, there is a simple analytical solution for the step index profile with $\Delta n \ll 1$ shown in Fig. (2.4 b) [54, p. 34]. The solution are linearly polarized transversal field distributions (the LP$_{lp}$ modes) with different propagation constants. They can be distinguished by their azimuthal order l and their radial order p. As long as $\underline{V} < 2.405$, the fiber is single-mode supporting only the fundamental mode LP$_{01}$. Its normalized propagation constant $\underline{B}(\underline{V})$ is implicitly given by the characteristic equation [55, p. 261], [53, p. 131], [54, p. 39]

$$\frac{\underline{V}\sqrt{1-\underline{B}}J_1(\underline{V}\sqrt{1-\underline{B}})}{J_0(\underline{V}\sqrt{1-\underline{B}})} - \frac{\underline{V}\sqrt{\underline{B}}K_1(\underline{V}\sqrt{\underline{B}})}{K_0(\underline{V}\sqrt{\underline{B}})} = 0. \quad (2.43)$$

with J_n and K_n the Bessel and the modified Hankel function of order n, respectively. With the knowledge of $\underline{B}(\underline{V})$ and Eq. 2.42, the propagation constant β is given by

$$\beta = k_0(\underline{B}(\underline{V})(n_{co} - n_{cl}) + n_{cl}). \quad (2.44)$$

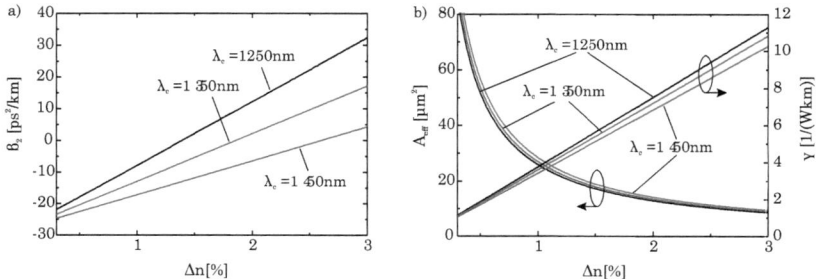

Figure 2.5: a) Fiber dispersion coefficient β_2 and b) effective mode area A_{eff} and nonlinear coefficient γ as a function of the relative index step Δn for different cut-off wavelengths λ_c (λ = 1550 nm, step-index profile)

Eq. 2.44 allows now to determine the dispersion coefficients β_n defined in Eq. 2.20. In particular, β_2 as a measure of the chromatic dispersion is given by

$$\beta_2 = \underbrace{\frac{n_{co}\Delta n}{\omega_0 c_0} \left.\frac{\underline{V}d^2(\underline{V}\cdot\underline{B})}{d\underline{V}^2}\right|_{\omega_0}}_{\text{Waveguide Dispersion}} + \underbrace{\frac{1}{c_0}\left[\omega_0\left.\frac{d^2 n_{cl}}{d\omega^2}\right|_{\omega_0} + 2\left.\frac{dn_{cl}}{d\omega}\right|_{\omega_0}\right]}_{\text{Material Dispersion}}. \qquad (2.45)$$

It comprises a term related to the material dispersion of silica and another term related to the waveguide dispersion. Since they have opposite signs, the waveguide dispersion can be used to cancel out the material dispersion [52]. Fig. (2.5 a) shows β_2 at the wavelength of 1550 nm as a function of the relative index step for different cut-off wavelengths λ_c which are defined as the wavelengths at which V = 2.405. The material dispersion was calculated using $n_{cl}(\omega)$ approximated by the Sellmeier equation [50, p. 6]. The graph shows that it is possible the reduce the chromatic dispersion to zero if Δn is increased to several percent.

Nonlinear coefficient As given in Eq. 2.40, the nonlinear coefficient depends on the nonlinear refractive index n_2 defined by Eq. 2.29 and the effective mode area A_{eff} defined by Eq. 2.33. The nonlinear refractive index of silica is a material constant slightly increasing with Δn due to the core doping with Germanium [50, p. 432]. Its main part stems from the electronic response of the material. The effective mode area is determined from the LP_{01} mode profile which can be approximated by a Gaussian distribution [55, p. 338]

$$F(x,y) = A_0\, e^{-\frac{x^2+y^2}{w^2}} \qquad (2.46)$$

where the mode radius may be defined by [55, p. 341]

$$w = \frac{d}{2\sqrt{\ln V}}. \tag{2.47}$$

Inserting the Gaussian mode profile in Eq. 2.33, the effective mode area is given by

$$A_{\text{eff}} = \pi w^2. \tag{2.48}$$

It is shown together with the resulting nonlinear coefficient in Fig. (2.5 b) as a function of the relative index step. If Δn is increased to several percent, it is possible to increase the nonlinear coefficient to $\gamma = 10/(\text{W km})$. Since this high Δn also supports low dispersion around 1550 nm as discussed above, it is the key for strong nonlinear interaction inside the HNLF. The first limitation of this approach is given by the single-mode cut-off frequency coming closer to 1550 nm when increasing γ while keeping the dispersion low at 1550 nm, as shown in Fig. (2.5). The second limitation are the losses that increase with Δn as shortly discussed in the following [52].

Absorption The loss of a silica fiber as a function of the wavelength shows a flat characteristic around 1550 nm justifying the assumption of a frequency-independent loss coefficient α_0. One can distinguish between different loss mechanisms [52],

$$\alpha_0 = \alpha_{\text{absorption}} + \alpha_{\text{scattering}} + \alpha_{\text{bending}}. \tag{2.49}$$

The absorption loss is due to electronic, molecular and color center material absorption. The scattering loss incorporates the attenuation due Rayleigh scattering and scattering due to waveguide imperfections such as defects or stress. The last term is the attenuation due to fiber bending. In a HNLF, $\alpha_{\text{scattering}}$ is typically increased in comparison to a SSMF since this type of loss increase nearly linearly with Δn due to the Germanium doping inside the core [52, 56].

2.2.3 Numerical Method

The NLS equation Eq. (2.39) is a nonlinear partial differential equation. Despite some special cases, analytical solutions can only be obtained using approximations. Therefore, one often relies on numerical solutions using the split-step Fourier method which is faster by up to two orders of magnitude compared with most finite-difference schemes [50, p. 41]. For this method, the NLS equation is formally rewritten in the form

$$\frac{\partial}{\partial z} A(z,t) = \left(\hat{D} + \hat{N} \right) A(z,t) \tag{2.50}$$

with the operators

$$\hat{D} = i \sum_{n=2}^{4} \frac{\beta_n}{n!}(i)^n \frac{\partial^n}{\partial T^n} - \frac{\alpha_0}{2}$$
$$\hat{N} = i\gamma |A(z,t)|^2 . \qquad (2.51)$$

Although in general, dispersion and nonlinearity act together, the split-step algorithm generates an approximate solution in assuming that dispersion and nonlinearity act independently when propagating the optical field over a small distance h_s. In this thesis, the software packet *ssprop* [57] was used to solve the NLS equation. It applies the symmetrized split-step scheme [50, p. 42], [58] where the solution to Eq. (2.50) is approximated by

$$A(z+h_s,T) \approx \exp(\frac{h_s}{2}\hat{D})\exp(\int_z^{z+h_s} \hat{N}(z')dz')\exp(\frac{h_s}{2}\hat{D})A(z,T). \qquad (2.52)$$

I.e., the propagation from z to $z+h_s$ is carried out in three parts. The first part is a step from z to $z+h_s/2$ where the dispersion acts alone and $\hat{N} = 0$. The second part only includes the nonlinearity, $\hat{D} = 0$, in a step from z to $z+h_s$. The third part is the step from $z+h_s/2$ to $z+h_s$ where the dispersion again acts alone and $\hat{N} = 0$. The exponential operator $\exp(h_s\hat{D}/2)$ is evaluated in Fourier space using the Fast-Fourier transform algorithm. The local error of the symmetrized split-step scheme is of the third order in the step size h_s [50, p. 42]. To further improve the accuracy, the integral in Eq. 2.52 is evaluated by using the trapezoidal rule,

$$\int_z^{z+h_s} \hat{N}(z')dz' \approx \frac{h_s}{2}\left[\hat{N}(z) + \hat{N}(z+h_s)\right]. \qquad (2.53)$$

Because $\hat{N}(z+h_s)$ is not yet known when solving the integral, it is necessary to follow an iterative procedure. Although this may be time-consuming, the overall computing time is reduced because the step size h can be increased due to the improved accuracy [50, p. 43].

2.2.4 Scattering Processes

Additionally to Rayleigh scattering, also Raman and Brillouin scattering occur in the fiber which are not included in Eq. 2.39. These are inelastic scattering processes that lead to an energy transfer from a pump wave to a frequency-downshifted probe wave that is also called Stokes wave where the frequency shift is material dependent. The energy difference is absorbed by the material in form of molecular vibrations for Raman scattering and in form of acoustic waves for Brillouin scattering. For intense pump waves, the nonlinear phenomena of stimulated Raman scattering (SRS) and stimulated

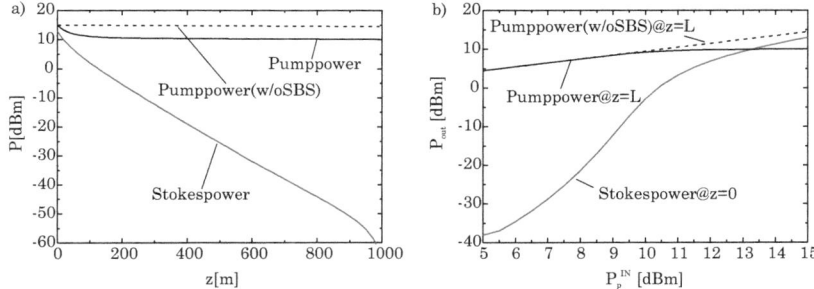

Figure 2.6: a) Pump and Stokes power as a function of the position inside the fiber in presence of SBS (for comparison, the pump power in absence of SBS is also given) and b) Pump and Stokes output power as a function of the pump input power in presence of SBS, for comparison, the pump output power in absence of SBS is also given (L = 1 km, γ = 10.3 /(W km), α_0 = 0.47 dB/km, $g_B(\Omega_B)/A_{eff}$ = 1.5 /(W m), λ = 1550 nm, T_0 = 300 K, Ω_B = 2 π 10 GHz, $\Delta\nu_B$ = 40 MHz)

Brillouin scattering (SBS) occur which lead to a rapidly growing Stokes wave such that most of the pump wave energy is transferred to it. In particular, SBS can severely limit the available optical power needed for nonlinear interactions in the HNLF as discussed below.

Stimulated Brillouin Scattering (SBS)

SBS generates a counterpropagating Stokes wave that is downshifted by the frequency shift $\Omega_B/(2\pi) \approx 10$ GHz in silica [50, p. 330]. For the case that the pump wave is a linearly polarized CW or quasi-CW signal and maintains its state of polarization along the fiber, the evolution of the powers of the pump and stokes wave can be described by two coupled differential equations [59],

$$\frac{dP_p}{dz} = -\int_{-\infty}^{\infty} \frac{g_B(\omega)}{A_{eff}} p_{st}(\omega) d\omega P_p - \alpha_0 P_p \qquad (2.54)$$

$$\frac{dp_{st}(\omega)}{dz} = -\frac{g_B(\omega)}{A_{eff}} (p_{st}(\omega) + p_{se}(\omega)) P_p + \alpha_0 p_{st}(\omega). \qquad (2.55)$$

Here, P_p is the pump power and p_{st} is the Stokes power spectral density with the Stokes power P_{st} given by

$$P_{st} = \int_{-\infty}^{\infty} p_{st}(\omega) d\omega. \qquad (2.56)$$

$p_{se}(\omega)$ is the power spectral density of the spontaneous Brillouin scattering given by

$$p_{se}(\omega) = \frac{2\pi k_B T_0}{h\Omega_B} \frac{h\omega}{2\pi}. \qquad (2.57)$$

where k_B, T_0, and h are the Boltzmann constant, the temperature and the Planck constant, respectively. The Brillouin gain spectrum $g_B(\omega)$ has a Lorentzian line shape [50, p. 331],

$$g_B(\omega) = g_B(\Omega_B) \frac{(\pi \Delta v_B)^2}{(\omega - \Omega_B)^2 + (\pi \Delta v_B)^2} \qquad (2.58)$$

with the Brillouin peak gain $g_B(\Omega_B) \approx 3-5 \times 10^{-11}$ m/W and the Brillouin gain bandwidth $\Delta v_B \approx 40$ MHz. Eqs. (2.54) can be solved applying the boundary conditions $P_p(z=0) = P_p^{IN}$ and $P_{st}(z=L) = 0$. The latter condition reflects the fact that no Stokes wave is inserted in the fiber but is initially generated by spontaneous Brillouin scattering. Fig. 2.6a shows the evolution of the pump and Stokes power along a fiber with the length of 1 km. The Stokes power increases exponentially along the fiber resulting in a depletion of the pump wave by 5 dB close to the fiber input. As depicted in 2.6b, an increase of the input pump power does not result in a higher output pump power but in a higher Stokes output power. The pump input power level, above which the pump output power saturates, is called the Brillouin threshold power. After a common definition, the threshold power is defined as the pump input power at which the Stokes output power equals the pump output power. Then, it is approximately given by [60]

$$P_{th,\text{SBS}} \approx \frac{21 A_{eff}}{g_B(\Omega_B) L}. \qquad (2.59)$$

Thus, the small A_{eff} in the HNLF decreases the threshold power in the same way as it increases the nonlinear coefficient. This puts a major limitation on the available pump power for nonlinear interactions.

Stimulated Raman Scattering (SRS)

For SRS, the Stokes wave can occur co- as well as counterpropagating with respect to the pump wave. This is called forward and backward SRS, respectively. The Raman gain spectrum for fused silica is with ≈ 30 THz wider by 3 orders of magnitude than the Brillouin gain spectrum with a peak at a Raman frequency Ω_R of about 13 THz [50, p. 276]. The peak gain $g_R(\Omega_R)$ is in the range of $1 \cdot 10^{-13}$ m/W and therefore 3 orders of magnitude lower than the Brillouin peak gain. The SRS threshold power, defined in a similar way as the SBS threshold power, is approximately given by [60]

$$P_{th,\text{SRS}} \approx \frac{16 A_{eff}}{g_R(\Omega_R) L}. \qquad (2.60)$$

For backward SRS, the numerical factor is 20 instead of 16. Due to the much lower peak gain, the SRS threshold power is larger by 3 orders of magnitude than the SBS threshold power for quasi-CW signals.

2.2.5 Nonideal Fiber Structure

The discussion of the fiber modes and the chromatic dispersion in section 2.2.1 assumes implicitly that the structure of the fiber is ideal, i.e. that the cross-section is perfectly circular and does not change over the fiber length. However, during fabrication and packaging, the circular symmetry and the uniformity over the length may be distorted. This causes random variations in dispersion and birefringence characteristics of the fiber which can be an issue for nonlinear processes like FWM as will be discussed later.

Variation of the Zero-Dispersion Wavelength

The small effective area of the HNLF is realized by a high relative index step Δn and a small core radius d/2. Therefore, already small, random variations in core radius during the fiber drawing translate into non-uniform characteristics of the HNLF. In particular, its dispersion characteristics depend strongly on the core radius (and therefore on variations of the normalized cut-off frequency $\underline{V}_c$) as can be estimated from Fig. 2.5. Thus, the zero-dispersion wavelength λ_{zd}, defined by $\beta_2(\lambda_{zd}) = 0$, can vary randomly over several nm over the length of a HNLF [61],[62].

Residual Fiber Birefringence

Another issue are small departures from the ideal circular symmetry of the fiber cross section that change randomly due to fluctuations in the core shape or due to stress. In this case, the mode degeneracy breaks and the propagation constant β as well as the group velocity $v_g = 1/\beta_1$ becomes slightly different for the modes polarized in the x and y directions. Such birefringence fluctuations induce polarization-mode dispersion (PMD) and randomize the state of polarization of any optical field propagating through the fiber [50, p. 408]. The PMD can be quantified by the PMD parameter D_p [50, p. 12].

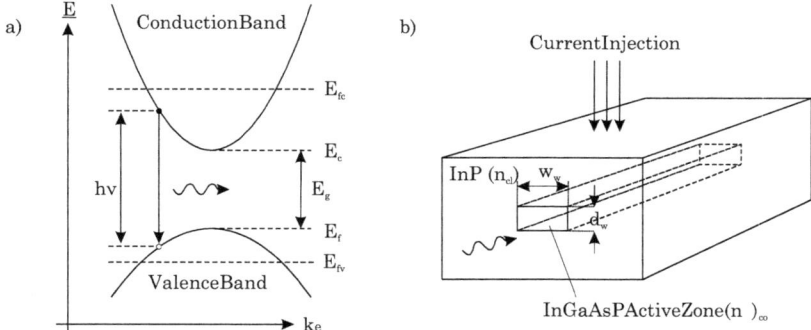

Figure 2.7: a) Simplified band diagram of a direct semiconductor and b) schematic structure of a SOA

2.3 Semiconductor Optical Amplifiers (SOA)

2.3.1 Structure

SOAs for the amplification of light with wavelengths between 900 and 1650 nm are based on the quaternary compound crystal Indium-Gallium-Arsenide-Phosphide ($In_{1-x}Ga_xAs_yP_{1-y}$). The mole fractions x and y denote to which amount Indium and Phosphide are replaced by Gallium and Arsenide, respectively. As long as $x = 0.4y + 0.067y^2$, the different compounds are lattice matched to InP allowing to fabricate complex SOA structures using epitaxial growth on InP substrates. Most importantly, $In_{1-x}Ga_xAs_yP_{1-y}$ is a direct semiconductor for all y if lattice matched. Fig. 2.7a depicts the simplified band diagram of such a crystal that shows the energy of the free carriers (electrons in the conduction band (CB), holes in the valence band (VB)) as a function of their wave vector magnitude k_e. Since it is a direct semiconductor, the energy minimum of the CB and the energy maximum of the VB are positioned at the same k_e. The minimum energy difference between CB and VB is called band gap energy and is dependent on the mole fractions x and y. By injecting current into the SOA, free carriers are generated in the CB and the VB. In the quasi-equilibrium, their distribution can be described by Fermi functions with separate Fermi energies E_{fc} and E_{fv} for the CB and the VB, respectively. If light with a frequency f (photon energy hf) is injected into the SOA, it is amplified by stimulated emission if

$$E_{\text{gap}} < hf < E_{\text{fc}} - E_{\text{fv}}. \tag{2.61}$$

For $hf < E_{gap}$, the material is essentially transparent, while for $hf > E_{fc} - E_{fv}$, the absorption of the light dominates. A typical SOA structure is shown in Fig. 2.7b. The main part is the $In_{1-x}Ga_xAs_yP_{1-y}$ waveguide core (also called the active zone). It has a lower bandgap energy and a higher refractive index than the surrounding InP forming a double-hetero structure. This allows to confine the injected carriers (by energy barriers) and the optical wave (by total refraction) at the same time. The carriers are supplied from the top and bottom electrodes (which will be referred to as pumping in the following), while the light is radiated from the side facets which carry an antireflection coating to suppress Fabry-Perot resonances.

2.3.2 Pulse Propagation Equation

As discussed above, due to the resonant nature of the nonlinearity of the SOA, Eq. 2.34 cannot be used. Instead, Eq. 2.22 together with Eq. 2.17 has to be used. For this purpose, the susceptibility can be divided into two parts [51, p. 26],

$$\chi(x,y,\omega) = \chi_0(x,y,\omega) + \chi_p(x,y,\omega). \tag{2.62}$$

χ_0 is the susceptibility in absence of pumping including the refractive index profile of the waveguide and the material absorption while χ_p takes into account the effect of the injected carriers. Then, the complex dielectric constant $\epsilon(x,y,\omega)$ is given by

$$\epsilon(x,y,\omega) = \underbrace{1 + \Re\{\chi_0(x,y,\omega)\}}_{\epsilon_b(x,y,\omega)} + \underbrace{i\Im\{\chi_0(x,y,\omega)\} + \chi_p(x,y,\omega)}_{\Delta\epsilon(x,y,\omega)}. \tag{2.63}$$

$\Delta\epsilon(x,y,\omega)$ is now inserted in Eq. 2.17 giving the perturbation of the propagation constant. Assuming that $\Im\{\chi_0\}$ and χ_p are constant in the active zone and zero outside, this yields

$$\Delta\beta(\omega) = -i\frac{\Gamma}{2}g(\omega)(1+i\alpha_H) + i\frac{a_{int}}{2}. \tag{2.64}$$

Γ is the confinement factor defined as

$$\Gamma = \frac{\int_{-d_w/2}^{d_w/2} \int_{-w_w/2}^{w_w/2} |F(x,y)|^2 dxdy}{\int_{-\infty}^{\infty} \int_{-\infty}^{\infty} |F(x,y)|^2 dxdy} \tag{2.65}$$

denoting the fraction of signal power within the core. Furthermore, the gain coefficient g was introduced as

$$g(\omega) = -\frac{k_0}{n_0}\Im\{\chi_0 + \chi_p\} \tag{2.66}$$

as well as the differentially formulated alpha factor [63]

$$\alpha_H = \frac{\partial \Re\{\chi_p\}/\partial N}{\partial \Im\{\chi_p\}/\partial N} \tag{2.67}$$

that describes the phase-amplitude coupling in the semiconductor material. N denotes the total carrier density in the SOA. The differential formulation of α_H allows to neglect any other non-carrier dependent real part in Eq. 2.64 which would lead to an additional constant phase shift. The (frequency independent) internal loss coefficient a_{int} was introduced phenomenologically and is related to loss due to scattering processes. Inserting $\Delta\beta(\omega)$ in Eq. 2.22 gives the pulse propagation equation for the SOA in the frequency domain,

$$\left(i\frac{\partial}{\partial z} + \beta_1(\omega-\omega_0) - i\frac{\Gamma}{2}g(\omega)(1+i\alpha_H) + i\frac{a_{int}}{2}\right)A(z,\omega-\omega_0) = 0. \tag{2.68}$$

where the unperturbed propagation constant $\beta(\omega) \cong \beta_0 + \beta_1(\omega-\omega_0)$ since the chromatic dispersion is usually negligible in SOAs due to the short length [64, 65, 66]. The nonlinearity in Eq. (2.68) is introduced by the material gain $g(\omega)$ which actually depends on $A(z,\omega-\omega_0)$ as described in the next section. Furthermore, in Eq. (2.68), a noise term is missing accounting for the spontaneous emission in the SOA. It is phenomenologically included by adding the noise field $A_{SE}(z,\omega)$ which will be also defined in the next section. The noise propagates in +z as well as in -z direction and significantly contributes to the gain saturation. Therefore, it is important to introduce A^+ and A^- representing the field envelopes propagating in +z and -z directions. With these modifications, Eq. (2.68) takes the form

$$\left(\pm i\frac{\partial}{\partial z} + \beta_1(\omega-\omega_0) - i\frac{\Gamma}{2}g(\omega)(1+i\alpha_H) + i\frac{a_{int}}{2}\right)A^\pm(z,\omega-\omega_0) = \pm i A_{SE}(z,\omega). \tag{2.69}$$

Finally, one has to keep in mind that there will be also noise in the orthogonal polarisation contributing to the gain saturation. Its propagation is also governed by Eq. 2.69 (with adjusted parameters if any polarisation dependency shall be taken into account). For notational simplicity, the following discussion will be carried out under the tacit assumption that both polarisations are considered and the SOA does not show any polarisation dependency.

Mode Profile The propagation constant β and the mode profile F(x,y) can be obtained by a similar procedure as described in Sec. 2.2.1 for the HNLF, i.e. Eq. (2.15) has to be solved using the proper refractive index profile $\epsilon_b(x,y,\omega)$. The rectangular waveguide shown in Fig. 2.7b can be approximated by a slab waveguide because its width w_w is much bigger than its height d_w. Then, a waveguide parameter similar to that in the silica fiber can be introduced,

$$\underline{V}_w = \frac{d_w}{2}k_0\sqrt{n_{co}^2 - n_{cl}^2} \tag{2.70}$$

and Eq. (2.15) can be solved analytically to obtain the propagation constants and the field distributions of the different modes [55, p. 240]. The single-mode condition is given by $\underline{V}_w < \pi/2$. If it is fulfilled only the fundamental modes TE_0 and TM_0 can propagate. The confinement factor of the TE_0 mode can be approximated by [51, p. 45]

$$\Gamma = \frac{\underline{V}_w^2}{0.5 + \underline{V}_w^2}. \tag{2.71}$$

The slightly lower confinement factor of the TM_0 mode in the slab waveguide would result in polarisation dependent gain. There are several ways to compensate for that using more elaborate waveguide structures [67, 68] so that the assumption of a polarization-independent SOA is justified.

2.3.3 Gain Modeling

The material gain g takes into account the resonant interaction of the photons with the carriers in the CB and the VB, i.e. the stimulated emission and absorption of photons in the SOA. The modeling of the material gain is generally challenging since it depends on the actual carrier distribution in the VB and the CB. One usually differentiates between inter- and intraband effects [69, 70, 71]. The former essentially describe the gain change due to a change of the total number of the carriers in the CB and the VB while they remain in the thermal (quasi-)equilibrium with the surrounding crystal lattice, i.e. they remain Fermi-distributed with lattice temperature. This is represented by the gain coefficient $g_{CDP}(N)$ (CDP - Carrier Density Pulsations). The latter incorporate all effects that change the material gain due to deviations from this thermal (quasi-)equilibrium distribution while the total number of carriers remains constant. These include carrier heating (CH), spectral hole burning (SHB), free carrier absorption (FCA) and two-photon absorption (TPA) that are all represented by their own gain coefficient. Furthermore, each inter- and intraband effect is related to an individual change in the refractive index which is taken into account by different alpha factors. Therefore, it should be kept in mind that

$$g(1 + i\alpha_H) \equiv \sum_X g_X(1 + i\alpha_{H,X}). \tag{2.72}$$

where X = {CDP, CH, SHB, FCA, TPA}.

Carrier distribution in the thermal (quasi-)equilibrium The carrier distribution in the CB in the thermal (quasi-)equilibrium and a parabolic

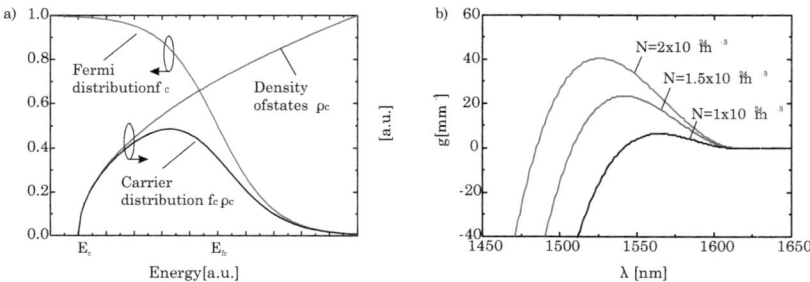

Figure 2.8: a) Schematic carrier distribution in the CB and b) material gain g_{CDP} as a function of the wavelength λ for different values of the total carrier density

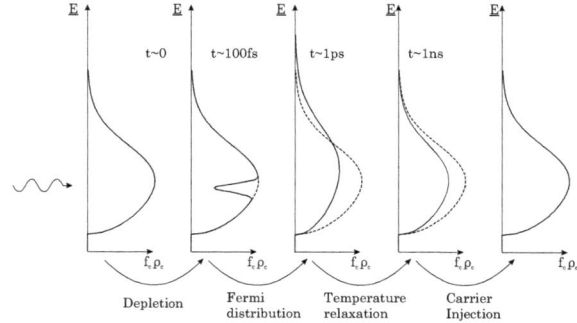

Figure 2.9: Intraband dynamics

band structure is schematically shown in Fig. 2.8a. It is the product of the density of states, ρ_c, and the Fermi distribution f_c defined by

$$f_c(\underline{E}) = \frac{1}{1+\exp\left(\frac{E-E_{f_c}}{k_B T_L}\right)} \tag{2.73}$$

For the parabolic band structure shown in Fig. 2.7a, $\rho_c \propto \sqrt{\underline{E}-E_c}$ for $\underline{E} > E_c$ and otherwise zero. The total carrier density N is given by

$$N = \frac{1}{V}\int_{E_c}^{\infty} \rho_c f_c d\underline{E}. \tag{2.74}$$

where V is the active zone volume.

Carrier dynamics Fig. 2.9 shows the dynamics of the carrier distribution for the case that a single short light pulse with a given wavelength is injected into the SOA. In the first step, its amplification depletes carriers at the energy level that participates in the transition and distorts the Fermi distribution. This depletion is called SHB and is quasi instantaneous. In the second step, the carriers return to a Fermi distribution by carrier-carrier scattering within ~100 fs. However, the carrier temperature is now increased which is called CH. In a similar way, FCA and TPA lead to deviations from the thermal (quasi-) equilibrium distribution by generating carriers at high energy levels and thereby increasing the carrier temperature. Thus, in the third step, the carrier temperature relaxes to the lattice temperature by carrier-phonon scattering within ~1 ps. In the last step, the original total carrier density is restored by carrier injection within ~1 ns. The resulting gain change is covered by g_{CDP}.

Material gain Since the material gain g is dependent on the carrier distribution in the CB and the VB, it is a function of the carrier density N and the wavelength λ. The peak gain is a sum over all gain contributions,

$$g_p = g_{CDP} + g_{CH} + g_{SHB} + g_{FCA} + g_{TPA}. \tag{2.75}$$

The contribution from CDP is given by

$$g_{CDP} = a_N(N - N_{tr}); \tag{2.76}$$

while the others will be defined in the next section. Here, a_N is the differential gain and N_{tr} is the transparency carrier density. For the frequency dependence of g_{CDP}, one often uses phenomenological models in which experimental results are parametrized using polynomial functions. Here, the model from [72] is adopted which presents a slight variation of [73]. The gain is given by

$$g(N,\omega) = \begin{cases} 3g_{p,2}\left(\frac{\omega-\omega_z}{\omega_z-\omega_{p,2}}\right)^2 + 2g_{p,3}\left(\frac{\omega-\omega_z}{\omega_z-\omega_{p,3}}\right)^3 & \omega > \omega_z \\ 0 & \omega < \omega_z \end{cases} \tag{2.77}$$

with

$$g_{p,2} = g_p + \bar{a}a_N N_{tr}\exp(-N/N_{tr})$$
$$g_{p,3} = g_p + a_N N_{tr}\exp(-N/N_{tr})$$
$$\omega_{p,2} = \omega_g + b_0(N - N_{tr}) + \bar{b}\omega_c\exp(-N/N_{tr})$$
$$\omega_{p,3} = \omega_g + b_0(N - N_{tr}) + \omega_c\exp(-N/N_{tr})$$
$$\omega_z = \omega_{z0} + z_0(N - N_{tr}).$$

All parameters are defined in Appendix G. Essentially, Eq. 2.77 is a combination of a quadratic and a cubic polynomial in ω. The exponential terms are used to smooth the gain function for low carrier densities and can be neglected for N well above N_{tr}. Then, the gain peak value given by g_p and its angular frequency $\omega_{p,2} \cong \omega_{p,3}$ are linear functions of N. Fig. 2.8b shows g as a function of the wavelength for different total carrier densities for the case that the gain contributions from the intraband effects are set to zero.

Noise The amplification of light using stimulated emission is inevitably connected to the generation of noise due to spontaneous emission. The ratio between the spontaneous and stimulated emission rate is given by the inversion factor n_{sp} [51, p. 226],

$$n_{sp}(\omega) = \frac{1}{1-\exp\left(\frac{\hbar\omega-(E_{fc}-E_{fv})}{k_B T_0}\right)}. \qquad (2.78)$$

Although the spontaneous emission spectrum is not constant, it is very broad in comparison to the gain spectrum. Therefore, it can be modeled as white noise with a constant noise spectral power density given by

$$\rho_{SE} = n_{sp}(\omega_0)\Gamma g_{CDP}\hbar\omega_0 \qquad (2.79)$$

and A_{SE} is defined in the time domain as a white Gaussian distributed noise process with the autocorrelation function

$$<A_{SE}(z,t)A_{SE}^*(z-z',t-t')> = \rho_{SE}\delta(z-z')\delta(t-t'). \qquad (2.80)$$

If a symmetric gain spectrum centered at ω_p with a bandwidth of $(E_{fc}-E_{fv})/\hbar - \omega_g n_{sp}(\omega_0)$ is assumed, and furthermore $\omega_0 \cong \omega_{p,2} \cong \omega_{p,3}$, $n_{sp}(\omega_0)$ can be approximated using Eq. 2.77 by noting that $\hbar\omega_0 - (E_{fc}-E_{fv}) \approx \hbar(\omega_g - \omega_{p,2}) = -\hbar b_0(N-N_{tr})$ for high carrier densities. Thus,

$$n_{sp}(\omega_0) \approx \left(1-\exp\left(\frac{-\hbar b_0(N-N_{tr})}{k_B T_0}\right)\right)^{-1}. \qquad (2.81)$$

The noise power added by the SOA per length Δz in the optical noise bandwidth B_N is given by

$$P_{SE} = \rho_{SE} B_N \Delta z. \qquad (2.82)$$

2.3.4 Time-Domain Modeling

The set of two nonlinear differential equations for the forward- and backward propagating waves given by Eq. 2.69 is usually solved using numerical methods. Here, this is done using the method of finite differences [74]. Fig. 2.10 depicts the principle. The SOA is divided into longitudinal sections with a length corresponding to

$$\Delta z = v_G \Delta t \qquad (2.83)$$

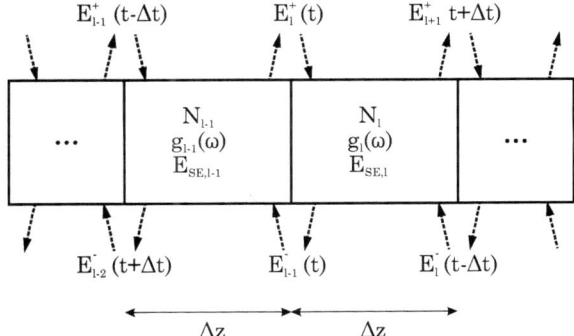

Figure 2.10: Principle of the finite-difference solver for the SOA pulse propagation equation

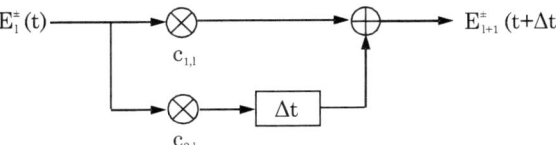

Figure 2.11: Schematic of the FIR filter used for the time-domain modeling

with Δt the sampling interval of the input signal. In each section denoted by the integer l, the total carrier density, the material gain and the spontaneous emission power are assumed to be constant. In each propagation step, two computational steps are performed: First, the local carrier density N_l, the local gain g_l (with its linear and nonlinear components) as well as the local nonlinear phase change and the local spontaneous emission field $A_{SE,l}^{\pm}$ are calculated. Second, the propagation equations Eq. (2.69) for the forward- and backward-traveling fields $A_l^{\pm}$ given at the interfaces of the sections are solved. The solution for the propagation through the segment l, i.e. over the distance Δz, is given by

$$A_{l\pm1}^{\pm}(\omega-\omega_0) = \overbrace{\exp\{\pm(\frac{\Gamma}{2}g_l(N_l,\omega)(1+i\alpha_H) - \frac{a_{int}}{2})\Delta z\}}^{G_l(N_l,\omega)}$$
$$\times \exp\{\pm i(\omega-\omega_0)\Delta t\}A_l^{\pm}(\omega-\omega_0) + A_{SE,l}(\omega). \qquad (2.84)$$

where it was assumed that all spontaneous emission power generated within the segment is added at its end.

It is generally preferable to solve Eq. (2.84) in the time domain since then all nonlinear interactions are taken into account automatically. To do this, $G_l(\omega)$ is approximated by the first-order finite impulse response (FIR) filter shown

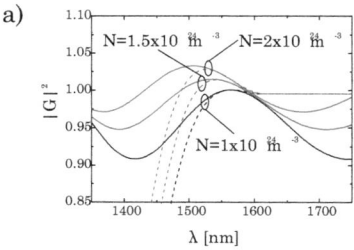 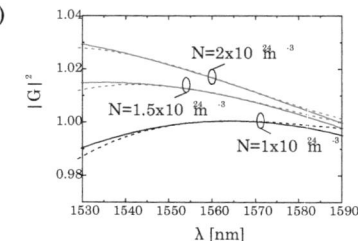

Figure 2.12: a) Power gain per section as a function of the wavelength by use of the FIR filter (straight lines) and calculated from the gain model Eq. 2.77 for different total carrier densities, b) same as a) but only for wavelength range of interest

in Fig. 2.11 which has the transfer function [74, 72]

$$G_l \approx G_{FIR,l}(\omega) = \left(c_{1,l} + c_{2,l} e^{i(\omega-\omega_0)\Delta t}\right) e^{i\Phi_l} \quad (2.85)$$

In each section, the coefficients $c_{1,l}$ and $c_{2,l}$ are adaptively fitted to the gain function Eq. 2.77 [72]. Their calculation is given in Appendix F. Φ_l includes the nonlinear phase change and is given by

$$\Phi_l = \phi_{CDP,l} + \sum_X \phi_{X,l}$$
$$\phi_{CDP,l} = \frac{1}{2}\Gamma\alpha_{H,CDP}(g_{CDP,l}(N_l) - g_{CDP}(N_{un}))\Delta z$$
$$\phi_{X,l} = \frac{1}{2}\Gamma\alpha_{H,X}g_{X,l}\Delta z$$

where X = {CH, SHB, FCA, TPA}. To keep Φ_l off from unnecessarily high values, the constant phase shift $-\Gamma\alpha_{H,CDP}g_{CDP}(N_{un}\Delta z$ was added here. As shown in Fig. 2.12a, the power transfer function $|G_{FIR,l}|^2$ of the first-order FIR filter has a sinusoidal spectral shape and does not match the power gain function $|G_l|^2$ for all wavelengths. However, Fig. 2.12b shows that the match is very well in the wavelength range of interest around 1550 nm. Now, the inverse Fourier transform can be applied to Eq. 2.84 giving the pulse propagation equation in the time domain,

$$A_{l\pm1}^{\pm}(t+\Delta t) = \left(c_{1,l}(t)A_l^{\pm}(z,t) + c_{2,l}(t)A_l^{\pm}(t-\Delta t)\right)e^{j\Phi_l(t)} + A_{SE,l}(t). \quad (2.86)$$

Using Eq. 2.82, the noise field $A_{SE,l}(t)$ is modeled by [75]

$$A_{SE,l}(t) = \sqrt{\frac{\rho_{SE,l}\Delta z}{\Delta t}} \frac{x_1 + ix_2}{\sqrt{2}} \quad (2.87)$$

where x_1 and x_2 are independent Gaussian distributed random numbers with zero mean and unit variance. $B_s = 1/\Delta t$ is the simulation bandwidth.

Finally, the carrier dynamics have to be included which is conveniently done in time domain by using rate equations. The dynamics of N (and therefore of g_{CDP}) are modeled by [74]

$$\frac{dN_l}{dt} = \frac{I_B}{qw_w d_w \Delta z} - R(N_l) - v_G(g \cdot S)_l + \frac{\Gamma_{TPA}}{\Gamma} v_G \beta_{TPA} S_l^2. \qquad (2.88)$$

The first term at the right-hand side describes the increase of N due to the pumping. Thereby, I_B is the pump current and q is the electron charge. The second terms takes into account the decrease due to spontaneous recombination. The third term represents the change of N by stimulated emission or absorption. The fourth term is related to the increase in total carrier density by TPA. Here, β_{TPA} is the two-photon absorption coefficient and $\Gamma_{TPA} > \Gamma$ is the TPA confinement factor taking into account the tighter confinement of the square of the intensity profile [76]. The product $(g \cdot S)_l$ is defined by

$$(g \cdot S)_l = g_l^+ S_l^+ + g_l^- S_l^-. \qquad (2.89)$$

The photon density $S_l^\pm$ is given by

$$S_l^\pm = \frac{1}{2k_p}(|A_{l,x}^\pm|^2 + |A_{l\pm1,x}^\pm|^2) + \frac{1}{2k}(|A_{l,y}^\pm|^2 + |A_{l\pm1,y}^\pm|^2) \qquad (2.90)$$

with

$$k_p = h\nu d_w w_w v_G / \Gamma. \qquad (2.91)$$

The indices x and y show explicitly that also the contribution of the orthogonal polarisation has to be taken into account. The effective gain coefficients $g_l^\pm(t)$ are defined by

$$\exp\left((\Gamma g_l^\pm(t + \Delta t) - a_{int})\Delta z\right) = \frac{S_{l\pm1}^\pm(t + \Delta t)}{S_l^\pm(t)}. \qquad (2.92)$$

The spontaneous recombination term has the form

$$R(N_l) = A_{nr} N_l + B_{sp} N_l^2 + C_{Auger} N_l^3 \qquad (2.93)$$

where the first term describes recombination at defect states, the second term spontaneous radiative recombination and the third term Auger recombination.
Similarly, the dynamics of the intraband processes are modeled by rate equa-

tions for their corresponding gain coefficient [71, 72],

$$\frac{\partial g_{\mathrm{CH},l}}{\partial t} = -\frac{g_{\mathrm{CH},l}}{\tau_{\mathrm{CH}}} - \frac{\epsilon_{\mathrm{CH}}}{\tau_{\mathrm{CH}}}(g \cdot S)_l \qquad (2.94)$$

$$\frac{\partial g_{\mathrm{FCA},l}}{\partial t} = -\frac{g_{\mathrm{FCA},l}}{\tau_{\mathrm{CH}}} - \frac{\epsilon_{\mathrm{FCA}}}{\tau_{\mathrm{CH}}}a_N N_l S_l \qquad (2.95)$$

$$\frac{\partial g_{\mathrm{TPA},l}}{\partial t} = -\frac{g_{\mathrm{TPA},l}}{\tau_{\mathrm{CH}}} - \frac{\epsilon_{\mathrm{TPA}}}{\tau_{\mathrm{CH}}}\frac{\Gamma_{\mathrm{TPA}}}{\Gamma}v_G\beta_{\mathrm{TPA}}S,l^2 \qquad (2.96)$$

$$\frac{\partial g_{\mathrm{SHB},l}}{\partial t} = -\frac{g_{\mathrm{SHB},l}}{\tau_{\mathrm{SHB}}} - \frac{\epsilon_{\mathrm{SHB}}}{\tau_{\mathrm{SHB}}}(g \cdot S)_l - \left(\frac{\partial g_{\mathrm{CH},l}}{\partial t} + \frac{\partial g_{\mathrm{FCA},l}}{\partial t} + \frac{\partial g_{\mathrm{TPA},l}}{\partial t} + \frac{\partial g_{\mathrm{CDP},l}}{\partial t}\right). \qquad (2.97)$$

Thereby, the phenomenological gain compression factors ϵ_X take into account the strength of the particular intraband effect and the phenomenological time constants τ_{CH} and τ_{SHB} govern over the relaxation dynamics.

Chapter 3

Phase-Modulation Formats

While in chapter 2 the models of the HNLF and the SOA have been introduced, this chapter is devoted to the transmission and reception of the phase modulated signals that will be passed through the nonlinear devices in chapter 4 and 5. The chapter starts with the description of signal constellations and the corresponding transmitters. Second, the different receiver architectures, in particular the direct and the coherent receiver, are introduced. Their discussion is the basis for the analysis of the signal degradations in the next chapters. Finally, semi-analytical formulas for the BER estimation in presence of different degradations are given that will be used later for the quantitative characterization of phase distortions in terms of signal-to-noise ratio penalties.

3.1 Constellations

3.1.1 Ideal Constellations

Fig. 3.1 shows constellations of different advanced modulation formats together with the number of bits per symbol [77, 78]. (D)BPSK with two constellation points in the inphase components of the electrical field enables the encoding of 1 bit/symbol. When using both the inphase and the quadrature component of the electrical field, the format is called (D)QPSK and enables the encoding of 2 bits/symbol, thus doubling the spectral efficiency. A further increase of the number of phase states leads to 8-PSK with 3 bits/symbol. To reach 4 bits/symbol, it is more convenient to use different phase states and at the same time different amplitude states. This leads to the 16-QAM format which provides a better noise tolerance than 16-PSK. Fig. 3.1 shows the square 16-QAM format which in turn performs slightly better than star 16-QAM [6, p. 185]. The individual symbols of the modulated signal are defined

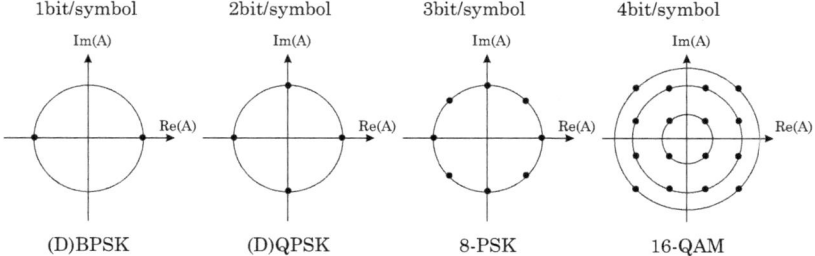

Figure 3.1: Constellations of several advanced modulation formats

by their electrical field,

$$A_k = \sqrt{P_k} e^{i\phi_k} a(t-t_k). \qquad (3.1)$$

where the allowed combinations for the symbol power P_k and the symbol phase ϕ_k depend on the modulation format and are shown in the constellations in Fig. 3.1. $a(t)$ is the pulse shape. The modulated signal is then given by

$$A_s(t) = \sum_k A_k = \sum_k \sqrt{P_k} e^{i\phi_k} a(t-t_k). \qquad (3.2)$$

Generally, all these modulation formats can be generated by modulating a CW laser signal using a single IQ modulator [79]. However, to avoid the use of multilevel electrical driving signals, more complex transmitter setups are proposed [6, Ch. 2].

3.1.2 Constellations in Presence of Noise

In practice, the generated and received signal constellations are never ideal due to the presence of noise. The two most important noise contributions always present in transmission systems are additive white Gaussian noise and laser phase noise. Thus, the modulated signal in the presence of noise is given by

$$\tilde{A}_s(t) = A_s(t) e^{i\phi_l(t)} + n_c(t). \qquad (3.3)$$

The laser phase noise $\phi_l(t)$ is generated in any laser diode (i.e. already in the transmitter laser diode) by spontaneous emission photons with random phase. The temporal evolution of the phase ϕ_l is a random walk where the random phase change within a time interval τ is given by [6, p. 16]

$$\Delta\phi_l(\tau) = \phi_l(t) - \phi_l(t-\tau). \qquad (3.4)$$

The phase difference $\Delta\phi_l(t)$ is Gaussian distributed with a variance of

$$<\Delta\phi_l^2(\tau)> = 2\pi\Delta\nu_l|\tau|. \qquad (3.5)$$

Thus, the laser phase noise is fully characterized by the laser linewidth $\Delta\nu_l$,

$$\phi_l(t) \equiv \phi_l(\Delta\nu_l). \tag{3.6}$$

The most important source of the complex additive white Gaussian (AWG) noise given by $n_c(t)$ are optical amplifiers present in the transmission channel that add amplified spontaneous emission noise. Other sources are quantum noise and the transmitter laser relative intensity noise. AWG noise can be characterized using the optical signal-to-noise ratio of a signal similarly defined as in Eq. 1.1,

$$\text{OSNR} = \frac{P_{av}}{2<n_c^2>|_{12.5\text{GHz}}} = \frac{P_{av}}{2\rho_{\text{AWG}}B_{\text{ref}}}.$$

Here, P_{av} is the average signal power. $<n_c^2>$ is the noise variance, i.e. the power contained in $n_c(t)$. The factor two refers to the fact that the noise is present in both polarizations. Since it is white noise, its noise power spectral density ρ_{AWG} is constant and a measurement bandwidth has to be chosen. For the OSNR, this measurement bandwidth B_{ref} is usually 12.5 GHz (or equivalently, 0.1 nm). Alternatively, one can choose the signal bandwidth, which corresponds ideally to the symbol rate R_s. Then, the signal-to-noise ratio

$$\text{SNR} = \frac{P_{av}}{<n_c^2>|_{R_s}} = \frac{P_{av}}{\rho_{\text{AWG}}R_s} = 2\text{OSNR}_s\frac{B_{ref}}{R_s} \tag{3.7}$$

may be defined [5, p. 67] which has the advantage that the related BER is independent of R_s. Note that for the SNR, only one noise polarization (that one parallel to the signal) is taken into account.

3.2 Reception

3.2.1 Direct Reception

In the direct receiver relying on a simple photodiode, only the intensity of the optical field can be detected. To evaluate phase modulation formats, differential detection of DPSK formats has to be used. This means that the information is encoded in optical phase differences of subsequent pulses allowing the use of an optical delay interferometer to convert the phase information into intensity information. In order to recover the original bit sequence after this operation, the transmitted bit sequence has to be precoded [6, p. 28]. Fig. 3.2a shows a direct, balanced receiver for DBPSK comprising a delay interferometer and a pair of photodiodes. Assuming that the received signal is corrupted

a)

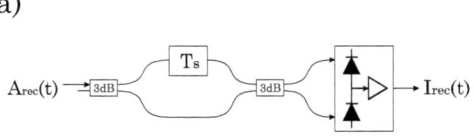

b)

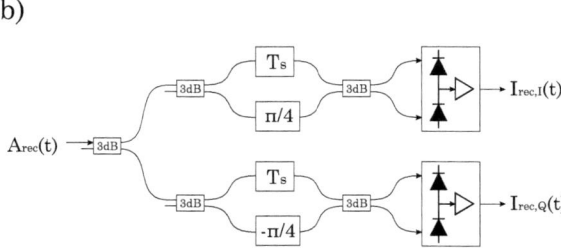

Figure 3.2: Receiver structures for direct detection of a) DBPSK and b) DQPSK [6, p. 71]

by noise but the pulse shapes $a(t-t_k)$ are not distorted [1], the received current is given by [6, p. 71]

$$I_{\text{rec}}(t) \propto \Re\{A_{\text{rec}}(t)A_{\text{rec}}^*(t+T_s)\}$$
$$\propto \sum_k \sqrt{\tilde{P}_k \tilde{P}_{k+1}} \tilde{a}^2(t-t_k)\cos(\Delta\tilde{\phi}_k) \qquad (3.8)$$

where $A_{\text{rec}}(t)$ is the received electrical field that may include AWG noise, laser phase noise and other distortions. Further, $a(t-t_k+T_s) = a(t-t_{k-1})$ was used. The phase difference $\Delta\tilde{\phi}_k$ includes the symbol phase difference $\Delta\phi_k = \phi_k - \phi_{k+1}$ but also any deterministic or statistical phase distortion, e.g. due to AWG noise or laser phase noise. Similarly, the received pulse powers $\tilde{P}_k$ include amplitude distortions. A decision error occurs if $\Delta\tilde{\phi}_k$ deviates more than ±90° from its ideal values 0° and 180°. As discussed in the next section, this may be due to linear or nonlinear noise contributions and/or deterministic distortions. Fig. 3.2b shows a configuration to receive DQPSK signals. Here, two DIs are used with ideal phases of 45° and −45°. The received currents are given by

$$I_{\text{rec,I}}(t) \propto \sum_k \sqrt{\tilde{P}_k \tilde{P}_{k+1}} \tilde{a}^2(t-t_k)\cos(\Delta\tilde{\phi}_k + \pi/4)$$
$$I_{\text{rec,Q}}(t) \propto \sum_k \sqrt{\tilde{P}_k \tilde{P}_{k+1}} \tilde{a}^2(t-t_k)\cos(\Delta\tilde{\phi}_k - \pi/4) \qquad (3.9)$$

---

[1]This condition is not fulfilled if any chromatic dispersion present during transmission is not compensated for leading to pulse shape broadening [50, pp. 53ff]

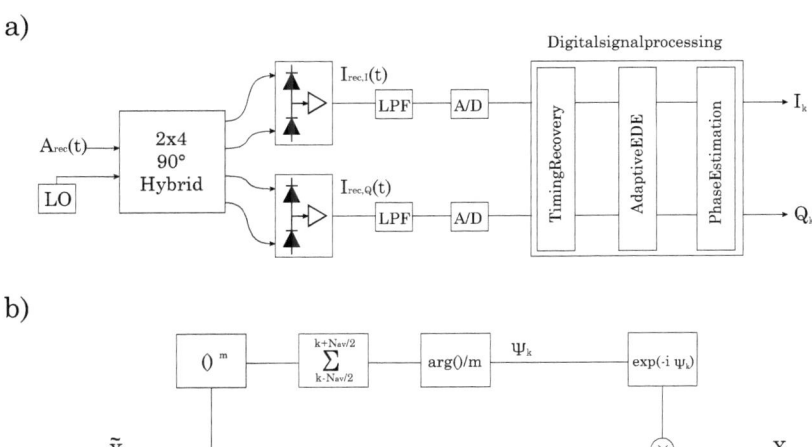

Figure 3.3: State-of-the-art coherent (intradyne) receiver for a single signal polarization

A correct symbol decision requires a correct decision on both $I_{\text{rec},I}$ and $I_{\text{rec},Q}$. A decision error occurs if $\Delta\tilde{\phi}_k$ deviates more than $\pm 45°$ from its ideal values $\{0°, 90°, 180°, 270°\}$. An equivalent option is to use a configuration with a 2×4 90° hybrid [6, p. 71].

In a similar way, also DPSK signals with order higher than 4 can be decoded. However, since the decoding effort increases strongly [6, pp. 68], these formats are not considered here.

3.2.2 Coherent Reception

Coherent detection relies on mixing of the signal wave with a local oscillator (LO) wave on a photodiode. The resulting photocurrent carries the full field information of the optical wave which allows the detection of both amplitude as well as phase modulation formats. For the latter, synchronous detection is possible which means that the absolute signal phase information is evaluated leading to a better noise performance compared to differential detection. However, this needs a phase synchronization of the input signal and the local oscillator. In traditional homodyne or heterodyne coherent receivers, the phase synchronization was achieved by locking the LO phase to the input signal phase by means of e.g. an optical phase-locked loop. Due to the recent progress in high-speed electronic signal processing, the phase synchronization is done today after detection by digital carrier synchronization techniques [78] allowing for a free-running, i.e. not phase locked LO wave. Such a receiver is called intradyne receiver.

Fig. 3.3a shows a typical state-of-the-art coherent receiver for single signal polarization [6, p. 100]. The extension to a polarization diversity receiver is straight forward. The receiver comprises of a local oscillator laser that is coupled together with the input signal light to a 2x4 90° hybrid. The four output signals are fed into two balanced receivers with two output currents given by [6, p. 93]

$$I_{rec,I}(t) \propto \sqrt{P_{LO}} \left(\sum_k \sqrt{\tilde{P}_k} \tilde{a}(t-t_k) \cos\left(\Delta\omega_{LO} t_k + \phi_{LO}(t_k) + \tilde{\phi}_k\right) + n_I(t_k) \right) \quad (3.10)$$

$$I_{rec,Q}(t) \propto \sqrt{P_{LO}} \left(\sum_k \sqrt{\tilde{P}_k} \tilde{a}(t-t_k) \sin\left(\Delta\omega_{LO} t_k + \phi_{LO}(t_k) + \tilde{\phi}_k\right) + n_Q(t_k) \right). \quad (3.11)$$

Thus, $I_{rec,I}$ and $I_{rec,Q}$ are proportional the two field quadratures of the received signal. $\Delta\omega_{LO}$ is the frequency offset between the received signal and the local oscillator and ϕ_{LO} represents the local oscillator phase noise. $\tilde{\phi}_k$ shall include the symbol phase ϕ_k but also additional phase distortions like the transmitter laser phase noise. However, it shall explicitly not contain the additive white Gaussian noise which is represented by n_I and n_Q in the in-phase and quadrature branch, respectively. For the pulse shape, $\tilde{a}$ is written denoting that the pulse shape may be distorted during transmission[2]. After low-pass filtering, the electrical current is digitized in an A/D converter. After timing recovery and electrical digital equalization to compensate for transmission distortions, the complex phasor after sampling is given by [6, p. 101]

$$\tilde{X}_k = I_{I,k} + iI_{Q,k} \propto \sqrt{P_{LO}} \left(\sqrt{\tilde{P}_k} a_k e^{i(\tilde{\phi}_k + \Delta\omega_{LO} t_k + \phi_{LO,k})} + n_c(t_k) \right) \quad (3.12)$$

The complex noise term is given by $n_c(t_k) = n_I(t_k) + in_Q(t_k)$. The local oscillator laser frequency offset and the local oscillator laser phase noise result in two additional phase distortion terms. The task of the phase estimation algorithm that follows the digital equalizer in Fig. 3.3a is to estimate and to remove the transmitter and local oscillator laser phase noise contained in $\tilde{\phi}_k$ and $\phi_{LO,k}$, respectively, as well as the local oscillator laser frequency offset $\Delta\omega_{LO}$. The goal is to recover the modulation information contained in ϕ_k. Due to its relatively simple implementation, one of the mostly used digital carrier phase estimation techniques is the feed forward m-th power scheme depicted in Fig. 3.3b [6, p. 102]. It can be applied to m-PSK, star QAM and square QAM

---
[2]For coherent reception, the pulse shape may be distorted e.g. by chromatic dispersion. This is electronically compensated for by digital equalization

formats (for the latter, an additional amplitude decision has to be made) [6, p. 105].

In this scheme, the complex phasor $\tilde{X}_k$ is first raised to the m-th power where m represents the number of phase states when applied to m-PSK formats. This removes the data phase modulation. In the second step, the phasor is averaged over N_{av} symbols to suppress the additive (zero-mean) Gaussian white noise contained in $n_c(k)$. In the last step, the phase is taken by an unwrapping arg-operation and the result is divided by m. Thus, the recovered carrier phase is given by

$$\Psi_k = \frac{1}{m}\arg\left(\sum_{l=k-(N_{av}-1)/2}^{k+(N_{av}-1)/2}(\tilde{X}_l)^m\right). \quad (3.13)$$

In the last step, the recovered carrier phase (containing the transmitter and local oscillator phase noise as well as the local oscillator frequency offset) is removed from the signal. Due to the m-th power operation, an m-fold phase ambiguity is induced. One way to solve this issue is differential encoding of the quadrants on the logical plane after phase estimation and data recovery [6, pp. 111].

3.3 Bit-Error Rate Estimation

3.3.1 Additive White Gaussian Noise

In the absence of other degradations, the fundamental limitation to the BER performance is additive white Gaussian (AWG) noise. A signal $\tilde{A}(t)$ may be distorted by the complex AWG noise contribution $n_c(t) = n_I(t) + in_Q(t)$. Then, each complex symbol $\tilde{A}_k$ of $\tilde{A}(t)$ has a PDF given by [80, p. 138]

$$\mathrm{PDF}_{\tilde{A}_k}(x+iy) = \frac{1}{2\pi\sigma_n^2}\exp\left(-\frac{(x-<\Re\{\tilde{A}_k\}>)^2+(y-<\Im\{\tilde{A}_k\}>)^2}{2\sigma_n^2}\right). \quad (3.14)$$

Here, $<\Re\{\tilde{A}_k\}>$ and $<\Im\{\tilde{A}_k\}>$ denote the mean real and imaginary signal part. Using Eq. 3.7, the noise variance in each signal quadrature is given by

$$\sigma_n^2 = <n_I^2(t)> = <n_Q^2(t)> = \frac{P_{av}}{(2\mathrm{SNR}_s)} \quad (3.15)$$

with the average signal power

$$P_{av} = <|\tilde{A}(t)|^2> \quad (3.16)$$

and the signal-to-noise ratio SNR_s. Under the assumption of an ideal receiver, the probability for a wrong decision on the symbol $\tilde{A}_k$ is given by

$$\mathrm{SER}_{\tilde{A}_k} = 1 - \iint_{F_k}\mathrm{PDF}_{\tilde{A}_k}(x+iy)dxdy \quad (3.17)$$

where the integration area F_k is defined by the decision thresholds that depend on the modulation format. The overall BER of the signal is then given by

$$\text{BER} \approx \frac{1}{\log_2(m)} <\text{SER}_{\tilde{A}_k}>\Big|_{\forall k} \qquad (3.18)$$

where the average is performed over all symbols. The formula takes into account Gray coded bit mapping which ensures that the closest neighbor constellation points differs only in a single bit independent on the number of bits per symbol. Then, by neglecting less likely transitions to non-closest neighbors, a symbol error results in only a single bit error. Gray codes can be used for both PSK and QAM formats [6, p. 23, p. 38].

For m-PSK formats, the decision variable is the phase. Because all symbols carry the same power and the PDF in Eq. 3.14 is symmetric, all symbol error rates are equal. Then, the BER for m-PSK signals is given by

$$\text{BER} = \frac{1}{\log_2(m)}\left(1 - \int_{-\pi/m}^{\pi/m} \text{PDF}_{\tilde{\phi}_k}(\phi)d\phi\right) \qquad (3.19)$$

with the PDF of the signal phase written as a Fourier series [80, 139],

$$\text{PDF}_{\tilde{\phi}_k}(\phi) = \frac{1}{2\pi} + \frac{1}{\pi}\sum_{l=1}^{\infty} c_l \cos(l\phi). \qquad (3.20)$$

The coefficients are given by [80, 139]

$$c_l = \frac{\sqrt{\pi \text{SNR}_s}}{2} e^{\text{SNR}_s/2}\left[I_{\frac{l-1}{2}}\left(\frac{\text{SNR}_s}{2}\right) + I_{\frac{l+1}{2}}\left(\frac{\text{SNR}_s}{2}\right)\right]. \qquad (3.21)$$

$I_k(x)$ is the k-th order modified Bessel function of the first kind. Eq. 3.20 can be evaluated analytically. The BER for m-PSK signals is given by

$$\text{BER}(\text{SNR}_s) = \frac{1}{\log_2(m)}\left(1 - \frac{1}{m} - \frac{2}{\pi}\sum_{l=1}^{\infty} \frac{c_l}{l}\sin\left(l\frac{\pi}{m}\right)\right). \qquad (3.22)$$

In difference to m-PSK formats, for m-DPSK formats, the decision variable is the phase difference $\Delta\tilde{\phi}_k = \tilde{\phi}_k(t) - \tilde{\phi}_k(t-T_s)$. $\tilde{\phi}_k(t)$ and $\tilde{\phi}_k(t-T_s)$ are two identical independently distributed random variables with a PDF given by Eq. 3.20. The differential phase $\Delta\tilde{\phi}_k$ has then a PDF of [80, p. 141]

$$\text{PDF}_{\Delta\tilde{\phi}_k}(\phi) = \frac{1}{2\pi} + \frac{1}{\pi}\sum_{l=1}^{\infty} c_l^2 \cos(l\phi) \qquad (3.23)$$

and the BER is given by

$$\text{BER}(\text{SNR}_s) = \frac{1}{\log_2(m)}\left(1 - \int_{-\pi/m}^{\pi/m} \text{PDF}_{\Delta\tilde{\phi}_k}(\phi)d\phi\right) \qquad (3.24)$$

$$= \frac{1}{\log_2(m)}\left(1 - \frac{1}{m} - \frac{2}{\pi}\sum_{l=1}^{\infty} \frac{c_l^2}{l}\sin\left(l\frac{\pi}{m}\right)\right). \qquad (3.25)$$

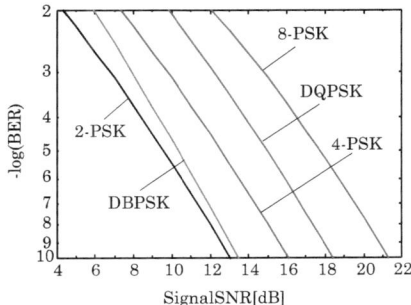

Figure 3.4: Back-to-back BER performance as a function of the signal SNR of different phase-shift keying formats.

Fig. 3.4 shows the (back-to-back) BER performance of different phase-shift keying formats calculated with Eqs. 3.22 and 3.25. Note that the BER is generally shown in this thesis as a function of the signal SNR[3].

3.3.2 Deterministic Phase Distortions

The correct reception of m-DPSK signals depends on the correct DI phase $\Delta\phi_{\text{DI}}$. A deviation θ_e from the optimum leads to a rotation of the signal constellation. In the presence of AWG noise, the PDF of the signal phase is then given by

$$\text{PDF}_{\Delta\tilde{\phi}_k}(\phi,\theta_e) = \frac{1}{2\pi} + \frac{1}{\pi}\sum_{l=1}^{\infty} c_l^2 \cos\left(l(\phi-\theta_e)\right). \qquad (3.26)$$

The BER is then given by [81], [80, p. 114]

$$\text{BER}(\text{SNR}_s,\theta_e) = \frac{1}{\log_2(m)}\left(1 - \frac{1}{m} - \frac{2}{\pi}\sum_{l=1}^{\infty}\frac{c_l^2}{l}\sin\left(l\frac{\pi}{m}\right)\cos(l\theta_e)\right). \qquad (3.27)$$

Of course, θ_e can also represent deterministic phase distortions of the signal accumulated during transmission. Similarly, also the constellation of coherently detected signals is rotated by phase distortions that are for some reason not removed by the CPE leading to a PDF in the presence of AWG noise of

$$\text{PDF}_{\tilde{\phi}_k}(\phi,\theta_e) = \frac{1}{2\pi} + \frac{1}{\pi}\sum_{l=1}^{\infty} c_l \cos\left(l(\phi-\theta_e)\right) \qquad (3.28)$$

and a BER of

$$\text{BER}(\text{SNR}_s,\theta_e) = \frac{1}{\log_2(m)}\left(1 - \frac{1}{m} - \frac{2}{\pi}\sum_{l=1}^{\infty}\frac{c_l}{l}\sin\left(l\frac{\pi}{m}\right)\cos(l\theta_e)\right). \qquad (3.29)$$

---

[3]In the literature, the BER is often shown as a function of the signal SNR per bit, i.e. $\text{SNR}_s/\log_2(m)$.

3.3.3 Nonlinear Phase Noise

A particular phase distortion is nonlinear phase noise. Its physical origin is discussed in the sections 2.1.3 and 2.1.3 while its impact will be discussed in sections 4.4, 5.3 and 5.4. Its phase PDF is to a good approximation a Gaussian distribution with variance σ_{nl}^2, [80, p. 178]

$$\text{PDF}_{\Delta\Phi_{nl}}(\phi) = \frac{1}{\sqrt{2\pi\sigma_{nl}^2}} \exp\left(-\frac{\phi^2}{2\sigma_{nl}^2}\right). \tag{3.30}$$

where it was assumed that $<\Delta\Phi_{nl}> = 0$. The overall PDF of the symbol phase degraded by AWG noise and nonlinear noise is the convolution of the two individual PDFs given by Eqs. 3.20 and 3.30,

$$\text{PDF}_{\tilde{\phi}_k}(\phi, \sigma_{nl}^2) = \frac{1}{2\pi} + \frac{1}{\pi}\sum_{l=1}^{\infty} c_l \cos(l\phi) e^{-\frac{l^2}{2}\sigma_{nl}^2}. \tag{3.31}$$

Insertion in Eq. 3.24 leads to the BER for m-PSK signals,

$$\text{BER}(\text{SNR}_s, \sigma_{nl}^2) = \frac{1}{\log_2(m)}\left(1 - \frac{1}{m} - \frac{2}{\pi}\sum_{l=1}^{\infty} \frac{c_l}{l}\sin\left(l\frac{\pi}{m}\right) e^{-\frac{l^2}{2}\sigma_{nl}^2}\right). \tag{3.32}$$

For m-DPSK, the decision variable is the phase difference. Then, the PDF and the BER are given by [82]

$$\text{PDF}_{\Delta\tilde{\phi}_k}(\phi, \sigma_{nl}^2) = \frac{1}{2\pi} + \frac{1}{\pi}\sum_{l=1}^{\infty} c_l^2 \cos(l\phi) e^{-l^2\sigma_{nl}^2}. \tag{3.33}$$

and

$$\text{BER}(\text{SNR}_s, \sigma_{nl}^2) = \frac{1}{\log_2(m)}\left(1 - \frac{1}{m} - \frac{2}{\pi}\sum_{l=1}^{\infty} \frac{c_l^2}{l}\sin\left(l\frac{\pi}{m}\right) e^{-l^2\sigma_{nl}^2}\right), \tag{3.34}$$

respectively.

Chapter 4

Parametric Amplifiers and Wavelength Converters based on Four-Wave Mixing in HNLF

After having introduced the model of the HNLF in chapter 2 and the transmission and reception of phase-modulated signals in chapter 3, this chapter is devoted to the analysis of phase distortions introduced by HNLF-based FOPAs. For this purpose, the characteristics of FOPAs are discussed in terms of the gain spectrum and the noise figure. Then, possible sources for phase distortions are identified using an analytical approximate solution of the FOPA equations derived from the model presented in chapter 2. In the following main part of this chapter, all identified phase distortions are discussed in detail and their individual impact on the BER of various directly and coherently detected phase-modulation formats is quantified using the BER formulas given in chapter 3.

4.1 General Characteristics

In the following section, only the ideal gain/conversion efficiency and noise figure of HNLF-based parametric wavelength converters can be discussed for brevity. One should notice that experimental results come very close to these theoretical curves as shown e.g. in [83, 84]. For a complete overview, possible deviations from the ideal values will be briefly mentioned. Another aspect that cannot be treated is polarization dependency. Generally, FWM is strongly polarization dependent so that polarization diversity schemes have to be applied. Because parametric amplification and wavelength conversion is a linear operation on the signal (as long as pump depletion is avoided), the

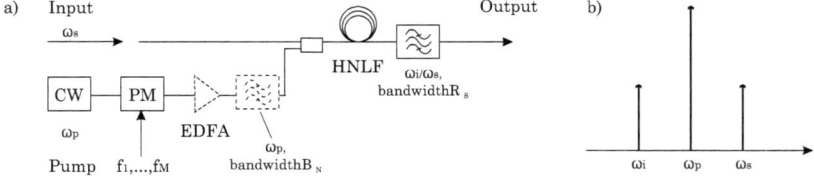

Figure 4.1: a) Single-pump configuration of the HNLF-based FOPA, b) schematic HNLF output spectrum

behavior is similar as for the polarization dependent AOWCs presented here to a first approximation.

4.1.1 Setup

As discussed in section 2.1.3, one can distinguish between degenerate and non-degenerate FWM depending on how many light waves are coupled by the process. From an application point of view, degenerate and non-degenerate FWM translate into three different configurations for possible AOWCs [85] which will be discussed in the following.

Single-Pump Configuration

The single-pump (SP) configuration relies on degenerate FWM [1]. The setup is depicted in Fig. 4.1a. The (weak) input signal is combined with a single (strong) pump wave and fed into the HNLF. As will be discussed in detail in section 4.1.4, the pump wave is phase modulated to suppress SBS in the HNLF. To reach high pump powers, subsequent amplification by an EDFA is often used followed by a narrow band filter to suppress the ASE noise. Because the amplification could be avoided if a high power CW laser as the pump source is available, the EDFA and the filter are shown using a dashed line. The schematic HNLF output spectrum is given in 4.1b. A single converted signal (called idler in the following) is generated by the degenerate FWM which is filtered out by a bandpass filter. The nonlinear process is characterized by an energy transfer from the pump to the signal and the idler. Thus, parametric amplification of the signal is possible in this scheme. With Eq. 2.36, the idler frequency is given by

$$\omega_i = 2\omega_p - \omega_s. \tag{4.1}$$

---

[1]This process is also called modulational interaction when occuring together with other FWM processes in the case of non-degenerate FWM [85].

This process is a phase conjugating process, i.e. the idler is a phase conjugated copy of the signal, as indicated by the negative sign before ω_s in Eq. 4.1. Similarly, the phase matching condition Eq. 2.37 requires

$$\Delta B_{sp} = B_s + B_i - 2B_p \approx 0. \tag{4.2}$$

Using the Taylor expansion given in Eq. 2.19 and choosing $\omega_0 = \omega_{zd}$, i.e. expanding around the zero-dispersion frequency, the linear phase mismatch can be expressed as

$$\Delta B_{sp} = \beta_3(\omega_{zd})(\omega_p - \omega_{zd})(\omega_s - \omega_p)^2 \tag{4.3}$$
$$+ \beta_4(\omega_{zd})/12(\omega_s - \omega_p)^2 \left[(\omega_s - \omega_p)^2 + 6(\omega_p - \omega_{zd})^2\right].$$

Note that $\beta_2(\omega_{zd}) = 0$ and $\omega_s - \omega_p = -(\omega_i - \omega_p)$. Any impact of the third-order dispersion can be eliminated by choosing $\omega_p \cong \omega_{zd}$. In this case, the phase matching condition is fulfilled over a wide bandwidth and broadband single-pump FWM is possible because the fourth-order dispersion is usually small in optical fibers. However, with Eq. 4.1, this choice means that the idler frequency depends on the signal frequency, $\omega_i \cong 2\omega_{zd} - \omega_s$, so that an arbitrary mapping of any input frequency to any output frequency is not possible. This problem can be circumvented by using dispersion flattened HNLFs with a very small β_3 [2] so that the pump frequency may deviate from the zero-dispersion frequency without increasing the phase mismatch significantly. It was shown that this allows for arbitrary mapping of the input and output frequencies over the C-band with low variances in the conversion efficiency [86].

Dual-Pump Configuration

This configuration relies on non-degenerate FWM. The setup is depicted in Fig. 4.2. In contrast to the single-pump configuration, the input signal is combined with two (strong) pump waves. The schematic HNLF output spectrum shows that now the output signal can be chosen of three different idler waves generated by three different FWM processes. The first process is called phase conjugation (PC) and is characterized by an energy transfer from the two pumps to the signal and the idler. As for the single-pump configuration, parametric amplification of the signal is possible. With Eq. 2.36, the idler frequency is given by

$$\omega_{i1} = \omega_{p1} + \omega_{p2} - \omega_s. \tag{4.4}$$

---

[2]This corresponds to a very small dispersion slope. The connection between β_3 and the dispersion slope is given in App. D.

As the name indicates, it is also a phase conjugating process. The phase matching condition Eq. 2.37 requires

$$\Delta B_{pc} = B_s + B_{i1} - B_{p1} - B_{p2} \approx 0. \tag{4.5}$$

Using again the Taylor expansion given in Eq. 2.19 and choosing $\omega_0 = \omega_{zd}$, the linear phase mismatch can be expressed as

$$\Delta B_{pc} = \beta_3(\omega_{zd})(\omega_a^{pc} - \omega_{zd})\left[(\omega_s - \omega_a^{pc})^2 - (\omega_{p1} - \omega_a^{pc})^2\right] \tag{4.6}$$
$$+ \beta_4(\omega_{zd})/12\left[(\omega_s - \omega_a^{pc})^2 - (\omega_{p1} - \omega_a^{pc})^2\right]$$
$$\times \left[(\omega_s - \omega_a^{pc})^2 + (\omega_{p1} - \omega_a^{pc})^2 + 6(\omega_a^{pc} - \omega_{zd})^2\right]. \tag{4.7}$$

where

$$\omega_a^{pc} = (\omega_{p1} + \omega_{p2})/2 \tag{4.8}$$

is the symmetry frequency of the PC process. Furthermore, $\omega_s - \omega_a^{pc} = -(\omega_{i1} - \omega_a^{pc})$ and $\omega_{p1} - \omega_a^{pc} = -(\omega_{p2} - \omega_a^{pc})$ were used. To eliminate the impact of the third-order dispersion, the symmetry frequency must be chosen to $\omega_a^{pc} \cong \omega_{zd}$, i.e. the pumps have to be placed symmetrically around the zero-dispersion frequency. This enables broadband FWM but prevents arbitrary input to output frequency mapping because with Eq. 4.4 follows $\omega_{i1} \approx 2\omega_{zd} - \omega_s$. As for the single-pump process, the use of dispersion-flattened HNLFs can alleviate this problem.

A second idler is generated by the process called frequency conversion (FC) [3] which is characterized by an energy transfer from pump 2 and the signal to pump 1 and the idler. Thus, unlike for the SP and the PC process, the signal is attenuated on the cost of the idler so that parametric amplification is not possible. The idler frequency is given by

$$\omega_{i2} = \omega_{p2} + \omega_s - \omega_{p1}. \tag{4.9}$$

Furthermore, FC is the only FWM process that is not phase conjugating. The phase matching condition Eq. 2.37 requires

$$\Delta B_{fc} = B_{i2} + B_{p1} - B_s - B_{p2} \approx 0. \tag{4.10}$$

Using the Taylor expansion given in Eq. 2.19 and choosing $\omega_0 = \omega_{zd}$, the linear phase mismatch can be expressed as

$$\Delta B_{fc} = \beta_3(\omega_{zd})(\omega_a^{fc} - \omega_{zd})\left[(\omega_s - \omega_a^{fc})^2 - (\omega_{p1} - \omega_a^{fc})^2\right] \tag{4.11}$$
$$+ \beta_4(\omega_{zd})/12\left[(\omega_s - \omega_a^{fc})^2 - (\omega_{p1} - \omega_a^{fc})^2\right]$$
$$\times \left[(\omega_s - \omega_a^{fc})^2 + (\omega_{p1} - \omega_a^{fc})^2 + 6(\omega_a^{fc} - \omega_{zd})^2\right]. \tag{4.12}$$

---
[3]This process is also called Bragg Scattering in the literature [85].

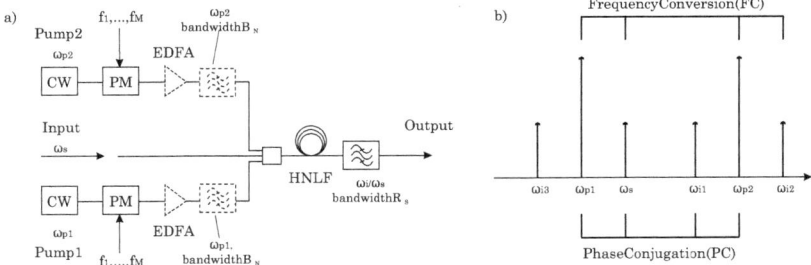

Figure 4.2: a) Dual-pump configuration of the HNLF-based FOPA, b) schematic HNLF output spectrum

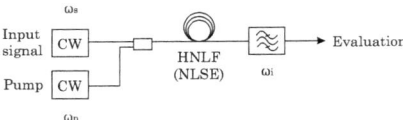

Figure 4.3: Single-pump simulation setup to characterize the conversion efficiency

where
$$\omega_a^{fc} = (\omega_s + \omega_{p2})/2 \qquad (4.13)$$
is the symmetry frequency of the FC process. Furthermore, $\omega_s - \omega_a^{fc} = -(\omega_{p2} - \omega_a^{fc})$ and $\omega_{i2} - \omega_a^{fc} = -(\omega_{p1} - \omega_a^{fc})$ were used. Also for the FC process, the symmetry frequency has to be chosen $\omega_a^{fc} \cong \omega_{zd}$ to eliminate the impact of the third-order dispersion. However, an arbitrary mapping of the input to the output frequency is nevertheless possible because $\omega_{i2} = 2\omega_{zd} - \omega_{p1}$ follows with Eq. 4.9, i.e. the idler frequency is not dependent on the signal frequency. This is as another unique feature of the FC process.

The third idler is generated by the same process as for the SP configuration. Here, it is characterized by an energy transfer of pump 1 to the signal and the idler.

4.1.2 Conversion Efficiency and Conversion Spectrum

The conversion efficiency G_i of any FOPA is the ratio of the output power of the converted signal and the input signal power, i.e. for the FWM-based FOPA it is given by
$$G_i = P_i(z=L)/P_s(z=0). \qquad (4.14)$$
Also the input signal exhibits amplification (or attenuation) due to the FWM process. The signal gain is defined as the output signal power to the input

signal power,
$$G_s = P_s(z=L)/P_s(z=0). \quad (4.15)$$

The conversion efficiency and the gain can be determined by a numerical simulation using Eq. 2.39 or by use of the approximate analytic solutions given in Appendix H. These solutions are obtained under quasi-CW conditions (i.e. by neglecting time derivatives) and do not take into account the depletion of the pumps by high signal and idler powers (i.e. the saturation effects) and the fiber loss. Nevertheless, they turn out to be very accurate over a wide wavelength range as discussed in the following. For simplicity, the same typical values for the HNLF parameters are used in all simulations and calculations unless otherwise stated. These parameters are summarized in Appendix E and, in addition, the corresponding dispersion as a function of the wavelength is shown in Fig. E.1.

Single Pump (SP)

Using the approximate solution of Appendix H, the conversion efficiency of the SP process is given by Eq. H.19 [85]

$$G_i^{sp} = \frac{\gamma^2 P_p^2}{g_{sp}^2} \sinh^2(g_{sp}L) = G_s^{sp} - 1 \quad (4.16)$$

with the gain coefficient g_{SP} given by Eq. H.16,

$$g_{sp}^2 = (\gamma P_p)^2 - \kappa_{sp}^2/4 \quad (4.17)$$

$$\kappa_{sp} = \Delta B_{sp} + 2\gamma P_p. \quad (4.18)$$

The phase mismatch parameter κ_{SP} consists of the linear phase mismatch ΔB_{sp} due to the chromatic dispersion and the nonlinear phase mismatch $2\gamma P_p$ due to XPM from the pump. The maximum conversion efficiency obtained for $\kappa_{SP} = 0$ is given by

$$\max\{G_i^{sp}\} = \sinh^2(\gamma P_p L) = \max\{G_s^{sp}\} - 1. \quad (4.19)$$

$\kappa_{SP} = 0$ means that the linear and the nonlinear part cancel out each other, $\Delta B_{sp} = -2\gamma P_p$. This is also called perfect phase matching and requires a negative linear phase mismatch. When neglecting β_4 in Eq. 4.3, perfect phase matching can be achieved if the pump is placed slightly below the zero-dispersion frequency. Fig. 4.4a shows the SP conversion efficiency for a CW input signal as a function of the wavelength detuning between the input signal wave and the pump wave. The detuning of the pump wavelength from the zero dispersion wavelength was varied from curve to curve which changes the linear phase mismatch. Only half of the spectrum is shown due to its

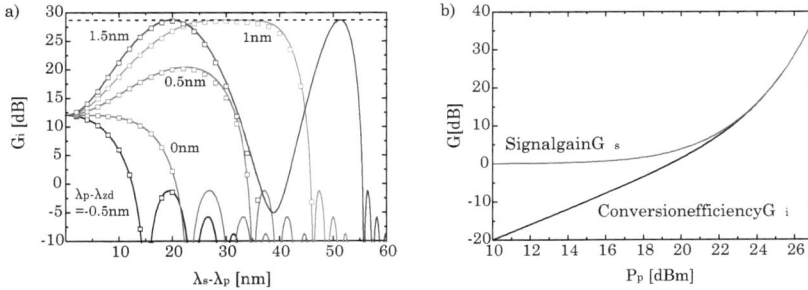

Figure 4.4: a) SP conversion efficiency G_i as a function of the pump-signal detuning $\lambda_s - \lambda_p$ for different detunings of the pump wavelength λ_p from the zero-dispersion wavelength λ_{zd} (solid lines - analytical solution using Eq. 4.16, symbols - numerical simulation based on Eq. 2.39). The parameters are given in the text. Only half of the conversion spectrum is shown due to its symmetric nature. b) SP signal gain G_s and conversion efficiency G_i as a function of the pump power P_p calculated with Eq. 4.19. The used parameters are given in the text.

symmetric nature. The symbols indicate results of the numerical simulation using Eq. 2.39 and the simulation setup given in Fig. 4.3 while the solid lines are given by the approximate solution using Eq. 4.16. The match is very good showing that the approximate solution is very accurate. The maximum gain obtained for perfect phase matching is indicated by the dashed line. The used parameters were L = 1 km, P_p = 26 dBm, P_s = -20 dBm, $\alpha = 0$, $\gamma = 10 (\text{W km})^{-1}$, zero-dispersion wavelength λ_{zd} = 1553nm, β_3 = 0.033ps^3/km and $\beta_4 = 2.5 \times 10^{-4}$ps^4/km. For $\lambda_p \leq \lambda_{zd}$, the conversion efficiency is a decreasing function of the pump-signal detuning since κ_{SP} is positive and increasing. In this case, the maximum is close to the pump where $\Delta B_{sp} = 0$ and the gain is quadratically dependent on the pump power. For $\lambda_p > \lambda_{zd}$, κ_{SP} is negative and increasing such that, at a given detuning, it cancels out the (always positive) nonlinear phase mismatch. Then, perfect phase matching is obtained leading to the maximum conversion efficiency. After that point, the linear phase mismatch increases further leading to a decreasing G_i. For a very large absolute value of the linear phase mismatch, $g_{sp}^2 < 0$ and therefore g_{sp} is purely imaginary. Then, the hyperbolic sine function in Eq. 4.16 gets an ordinary sine function and the conversion efficiency is a periodic function of the pump-signal detuning.

Fig. 4.4b shows the maximum signal gain and the maximum conversion efficiency calculated with Eq. 4.19. The used parameters were L = 1 km and

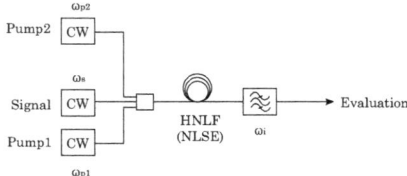

Figure 4.5: Dual-pump simulation setup to characterize the conversion efficiency

$\gamma = 10 (\text{W km})^{-1}$. One can see that the single-pump FOPA can provide parametric amplification for both the input signal and the idler if the pump power is high enough. However, the conversion spectrum (as well as the gain spectrum) is highly nonuniform since, as discussed above, G_i (and G_s) depend quadratically on the pump power close to the pump wavelength while they depend exponentially on the pump power at its maximum. A uniform gain and conversion spectrum can only be obtained by sacrifying gain and avoiding perfect phase matching by placing the pump exactly at the zero-dispersion wavelength.

Phase Conjugation (PC)

Using the approximate solution of Appendix H, the conversion efficiency of the PC process is given by Eq. H.44 [85],

$$G_i^{pc} = \frac{4\gamma^2 P_{p1} P_{p2}}{g_{pc}^2} \sinh^2(g_{pc}L) = G_s^{pc} - 1 \qquad (4.20)$$

with the gain coefficient g_{pc} given by Eq. H.41,

$$g_{pc}^2 = 4\gamma^2 P_{p1} P_{p2} - \kappa_{pc}^2/4 \qquad (4.21)$$

$$\kappa_{pc} = \Delta B_{pc} + \gamma(P_{p1} + P_{p2}). \qquad (4.22)$$

As for SP, the conversion efficiency is maximal if $\kappa_{pc} = 0$, i.e. for perfect phase matching. For PC, it is given by

$$\max\{G_i^{pc}\} = \sinh^2(2\gamma\sqrt{P_{p1}P_{p2}}L) = \max\{G_s^{pc}\} - 1. \qquad (4.23)$$

Similarly to the SP process, perfect phase matching can only be achieved for a negative linear phase mismatch. When neglecting β_4 in Eq. 4.6, one can see that $B_{pc} < 0$ can be realized by placing the symmetry wavelength ω_a^{pc} slightly above the zero-dispersion frequency. Fig. 4.6a shows the PC conversion efficiency for a CW input signal as a function of the wavelength detuning between the input signal wave and the symmetry wavelength $\lambda_a^{pc} = 2\pi c/\omega_a^{pc}$. Here, the

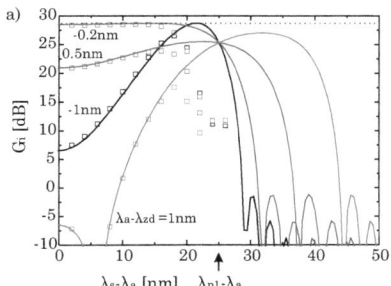
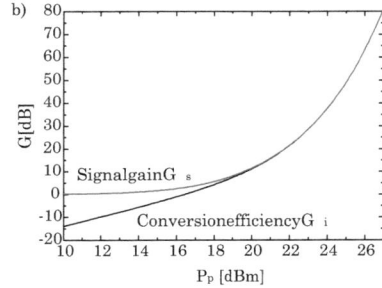

Figure 4.6: a) PC conversion efficiency G_i as a function of the signal wavelength λ_a detuning from the symmetry wavelength λ_a (defined by Eq. 4.8) for different detunings of λ_a from the zero-dispersion wavelength λ_{zd} (solid lines - analytical solution after Eq. 4.20, symbols - numerical simulation based on Eq. 2.39, b) PC signal gain G_s and conversion efficiency G_i as a function of the pump power P_p calculated by Eq. 4.23. The used parameters are given in the text.

detuning of the symmetry wavelength from the zero dispersion wavelength was varied from curve to curve to change the linear phase mismatch and only half of the spectrum is shown due to its symmetric nature. The pump wavelength is indicated at the x-axis. The pump separation was about 50 nm. The symbols indicate results of the numerical simulation using Eq. 2.39 and the simulation setup depicted in Fig. 4.5 while the solid lines are given by the approximate solution using Eq. 4.20. The maximum gain obtained for perfect phase matching is indicated by the dotted line. The used parameters were L = 1 km, $P_{p1} = P_{p2} = 23$ dBm, $P_s = -20$ dBm, $\alpha = 0$, $\gamma = 10(\text{W km})^{-1}$, zero-dispersion wavelength $\lambda_{zd} = 1553$nm, $\beta_3 = 0.033\text{ps}^3/\text{km}$ and $\beta_4 = 2.5 \times 10^{-4}\text{ps}^4/\text{km}$. The match between the simulation and the approximate solution is again very good except for signal wavelengths close to the pump wavelength. In this wavelength range, also the SP processes between the input signal and pump1 and between the PC generated idler and pump 2 are nearly phase matched which is not taken into account by the approximate solution. These interactions can be included by a four-side band analysis [85].

For the PC process, one can find an optimum detuning of the pump from the zero-dispersion wavelength that leads to a very flat conversion spectrum with maximum gain between the two pumps. At this detuning, the linear and nonlinear phase mismatch compensate each other over a wide wavelength range due to the presence of third and fourth order dispersion.

The maximum conversion efficiency (and the maximum gain) is shown in Fig.

4.6b as a function of the pump power calculated by Eq. 4.23. The used parameters were L = 1 km and $\gamma = 10(\text{W km})^{-1}$. To reach the same maximum gain as for SP, only half of the pump power (per pump) is needed. For PC, the signal gain is also given by $G_s = 1 + G_i$. In contrast to the SP process, a dual-pump FOPA using the PC process can provide the maximum conversion efficiency (and gain) together with a highly uniform conversion spectrum, i.e. together with the largest spectral width.

Frequency Conversion (FC)

Using the approximate solution of Appendix H, the conversion efficiency of the FC process is given by Eq. H.62 [85],

$$G_i^{fc} = \frac{4\gamma^2 P_{p1} P_{p2}}{g_{fc}^2} \sin^2(g_{fc} L) = 1 - G_s^{fc} \tag{4.24}$$

with the gain coefficient g_{fc} given by Eq. H.59,

$$g_{fc}^2 = 4\gamma^2 P_{p1} P_{p2} + \kappa_{fc}^2 / 4 \tag{4.25}$$

$$\kappa_{fc} = \Delta\beta + \gamma(P_{p2} - P_{p1}). \tag{4.26}$$

As mentioned above, the FC process does not lead to conversion efficiencies above unity. The maximum conversion efficiency occurs if $\kappa_{fc} = 0$ which means $\Delta B_{fc} = 0$ for equal pump powers. It is given by

$$\max\{G_i^{fc}\} = \sin^2(2\gamma\sqrt{P_{p1} P_{p2}} L) = 1 - \max\{G_s^{fc}\}. \tag{4.27}$$

Eq. 4.11 shows that $\Delta B_{fc} = 0$ can be achieved by placing the symmetry wavelength ω_a^{fc} slightly below the zero-dispersion frequency. In this case, $\omega_a^{fc} - \omega_{zd} \cong 0$ and

$$\Delta B_{fc} \approx \beta_3(\omega_{zd})(\omega_a^{fc} - \omega_{zd}) \left[(\omega_s - \omega_a^{fc})^2 - (\omega_{p1} - \omega_a^{fc})^2 \right] \tag{4.28}$$
$$+ \beta_4(\omega_{zd})/12 \left[(\omega_s - \omega_a^{fc})^4 - (\omega_{p1} - \omega_a^{fc})^4 \right],$$

i.e. the terms proportional to β_3 and β_4 have opposite signs and cancel out each other if β_3 and β_4 have the same sign.

Fig. 4.7a shows the FC conversion efficiency for a CW input signal as a function of the wavelength detuning between the input signal wave and the symmetry wavelength λ_a (defined by Eq. 4.13). Also here, the detuning of the symmetry wavelength from the zero dispersion wavelength was varied from curve to curve to change the linear phase mismatch and only half of the spectrum is shown due to its symmetric nature. The pump wavelength is indicated at the x-axis, the pump separation was about 50 nm. The symbols indicate

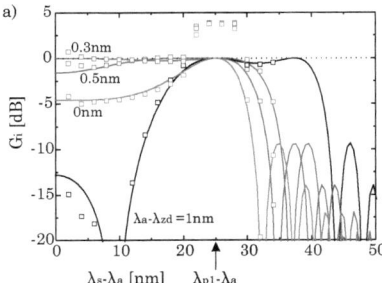 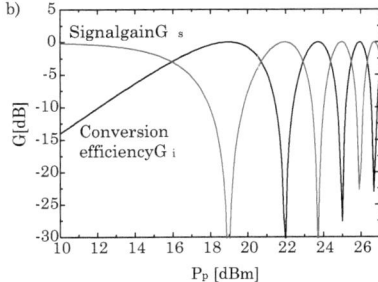

Figure 4.7: a) FC conversion efficiency G_i as a function of the signal wavelength λ_s detuning from the symmetry wavelength λ_a (defined by Eq. 4.13) for different detunings of the pump wavelength λ_p from λ_a (solid lines - analytical solution using Eq. 4.27, symbols - numerical simulation based on Eq. 2.39, b) FC signal gain G_s and conversion efficiency G_i as a function of the pump power P_p calculated by Eq. 4.27. The used parameters

results of the numerical simulation using Eq. 2.39 and the simulation setup shown in Fig. 4.5 while the solid lines are given by the approximate solution using Eq. 4.27. The maximum gain is indicated by the dashed line. The used parameters were L = 1 km, $P_{p1} = P_{p2}$ = 19 dBm, P_s = -20 dBm, $\alpha = 0$, $\gamma = 10(\text{W km})^{-1}$, zero-dispersion wavelength λ_{zd} = 1553nm, $\beta_3 = 0.033\text{ps}^3/\text{km}$ and $\beta_4 = 2.5 \times 10^{-4}\text{ps}^4/\text{km}$. The match between the simulation and the approximate solution is again very good except for signal wavelengths close to the pump wavelengths for the same reasons as for the PC process.

As for the PC process, one can find an optimum detuning of the pump from the zero-dispersion wavelength that leads to a very flat conversion spectrum with maximum gain. At this detuning, the linear phase mismatch is close to zero over a wide wavelength range since third and fourth order dispersion compensate each other. This shows that the dual-pump FOPA using the FC process can provide conversion efficiencies up to unity with a highly uniform conversion spectrum.

The maximum conversion efficiency is shown in Fig. 4.7b together with the signal gain as a function of the pump power calculated by Eq. 4.27. The used parameters were L = 1 km and $\gamma = 10(\text{W km})^{-1}$. Due to the sine function, $G_{i,max}^{FC}$ is periodic and never grows above unity. For FC, the signal gain is given by $G_s = 1 - G_i$ since every photon which is added to the idler is subtracted from the signal wave.

Degrading effects

In the simulations and analytical calculations in this section, only chromatic dispersion and the Kerr nonlinearity were taken into account and the HNLF was assumed to have an ideal uniform structure. However, as discussed in sections 2.2.4 and 2.2.5, in a real HNLF SBS and SRS can occur and the structure can be nonuniform. These effects degrade the magnitude and the uniformity of the conversion efficiency in particular of the dual-pump configuration which will be discussed shortly in the following.

SBS limits the maximum input power into the fiber. In order to achieve high conversion efficiencies, the pumps have to be phase modulated to suppress SBS. This will be discussed in detail in section 4.1.4.

SRS affects mainly the conversion efficiency of the dual-pump configuration by Raman-induced power transfer among the participating waves from shorter to longer wavelengths. Since the two pumps cannot maintain equal power levels along the fiber, the conversion efficiency is reduced even though the total power remains constant. In practice, the power of the high-frequency pump is chosen to be larger than that of the low-frequency pump at the HNLF input. The shape of the gain spectrum is not affected since the phase matching depends on the total power of the two pumps which is conserved in the undepleted case [50, pp. 398]. Spontaneous Raman scattering also affects the noise figure of single- and dual-pump AOWCs as discussed in the next section.

Zero-dispersion wavelength fluctuations limit the usable bandwidth of the FOPA, in particular of the dual-pump configuration. As shown in Fig. 4.6a, the shape of the conversion spectrum is strongly dependent on the position of λ_{zd} so that any fluctuation of λ_{zd} will reduce the uniformity [50, pp. 398]. This can be counteracted by reducing the pump spacing or by equalizing the fluctuations by intentionally applied longitudinal strain [87]. The latter has the advantage that also the SBS threshold can be increased.

Random birefringence randomizes the SOP of any optical field propagating through the HNLF and induces PMD effects. The former only reduces the average conversion efficiency by randomly changing the SOPs of the participating waves but keeping the relative SOPs constant. The latter also changes the relative orientation and distorts the uniformity of the spectrum. This can be avoided by using short lengths of low-PMD fibers and reduced wavelength spacings [50, pp. 410].

Although all these effects degrade the conversion spectrum of the FOPA, they do not introduce time-dependent distortions on short time scales, i.e. in the order of the bit period. Therefore, they will be neglected in the following sections.

4.1.3 Noise Figure

Single- and dual-pump AOWCs reduce the OSNR of the converted and the original signal. This can be described by the noise figure that is defined as [88]

$$\text{NF}_j = \frac{\text{OSNR}_{s,in}}{\text{OSNR}_{j,out}} = \frac{\text{SNR}_{s,in}}{\text{SNR}_{j,out}} \qquad (4.29)$$

where j = s,i (for signal and idler, respectively). Thereby, the input signal shall be limited by complex, Gaussian distributed quantum noise with a noise power spectral density of

$$\rho_{QN} = \frac{\hbar\omega}{2}. \qquad (4.30)$$

For the single-pump and the phase-conjugation process, this quantum noise spectral density will be amplified by an average gain equal to the signal gain $G_s^{sp/pc}$. Furthermore, FWM will also copy the noise at the idler frequency, adding additional noise to the signal. The copied noise will experience the conversion efficiency $G_i^{sp/pc} = G_s^{sp/pc} - 1$ (given in Eq. 4.16 for the single-pump process and in Eq. 4.20 for the phase-conjugation process). Adding the two contributions, the amplified quantum-noise power spectral density is given by

$$\rho_{AQN}^{sp/pc} = \frac{\hbar\omega}{2}(2G_s^{sp/pc} - 1). \qquad (4.31)$$

Then, the ideal noise figure for the amplified signal in the single-pump and in the phase-conjugation process can be written as [84, 89, 90]

$$\text{NF}_s^{sp/pc} = \frac{P_{av}/(R_s\hbar\omega/2)}{G_s^{sp/pc}P_{av}/((2G_s^{sp/pc}-1)R_s\hbar\omega/2)} = 2 - \frac{1}{G_s^{sp/pc}}. \qquad (4.32)$$

This is same noise figure as for an ideal EDFA. Similarly, the ideal noise figure for the idler in the single-pump and in the phase-conjugation process can be written as [84, 89, 90]

$$\text{NF}_i^{sp/pc} = \frac{P_{av}/(R_s\hbar\omega/2)}{G_i^{sp/pc}P_{av}/((2G_s^{sp/pc}-1)R_s\hbar\omega/2)} = 2 + \frac{1}{G_i^{sp/pc}}. \qquad (4.33)$$

For the frequency-conversion process, the conversion efficiency is given by Eq. 4.24, $G_i^{fc} = 1 - G_s^{fc}$, so that the output noise spectral density equals the input noise spectral density. The ideal noise figures for the frequency-conversion process are then given by [90]

$$\text{NF}_j^{fc} = 1/G_j^{fc}; \qquad (4.34)$$

with j = s,i. This is the noise figure of a passive device with loss $1/G_j^{fc}$.

The ideal noise figures are shown in Fig. 4.8 as a function of the maximum conversion efficiency $\max\{G_i\} \cong \max\{G_s\}$. For low conversion efficiencies, the

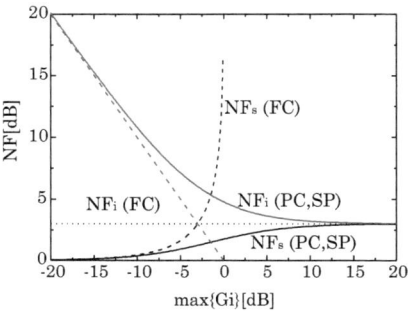

Figure 4.8: Ideal signal and idler noise figures NF_s and NF_i as a function of the maximum conversion efficiency $\max\{G_s\}$ for SP, PC and FC

idler noise figure of single-pump and phase-conjugation based FOPAs equals the reciprocal of the conversion efficiency while the signal noise figure is close to zero (with a gain close to 1). For high conversion efficiencies (and gains), they act like amplifiers with a noise figure approaching 3 dB. As mentioned above, the frequency-conversion based FOPA behaves always like a passive device. The signal noise figure approaches infinity for a unity conversion efficiency since the signal gain is zero at this point, i.e. the signal is completely suppressed. Meanwhile, the idler noise figure approaches 1, i.e., the SNR is not degraded. Fig. 4.8 shows the importance of a high conversion efficiency to build FOPAs with low noise figures.

The ideal noise figure of FWM-based FOPAs is degraded in practice by several effects. First, excess noise due to loss or gain caused by the Raman effect has to be included [91]. This explains why the lowest NF measured so far is in the order of 3.7 dB [10]. Furthermore, the noise figure is degraded by interaction with secondary idlers [92]. Third, also residual ASE noise at the signal and idler frequency stemming from the amplification of the pump(s) with the EDFA can increase the noise figure. Its suppression requires tight optical filtering of the pumps after the EDFA. Finally, the overall noise figure of a practical FOPA also includes the input and output HNLF coupling efficiency, the loss of the pump coupler and the loss of the bandpass filter. For high conversion efficiencies, only the input coupling loss and the pump coupler loss has an impact. The former is typically very low since the HNLF can be spliced to a SSMF with low losses and the latter can be minimized by using WDM couplers with low insertion loss.

In the literature, also excess noise due to time-dependent variations of the conversion efficiency induced by pump power fluctuations has been included in the FOPA noise figure [84, 89]. However, since the probability distribution function of the noise contribution is not Gaussian [93] it is not included here in the noise figure but it will be treated separately as nonlinear noise in sections 4.4.3 and 4.4.4

4.1.4 Suppression of SBS by Pump Phase Modulation

The maximum conversion efficiency of the SP process, i.e. of the single-pump configuration, is growing with the pump power as given by Eq. 4.19 and shown in Fig. 4.4. However, as discussed in section 2.2.4, the maximal power of a CW signal propagating through the HNLF is given by the SBS threshold power P_{th} given in Eq. 2.59 (here defined as the CW input power at which the transmitted power is equal to the reflected power). Inserting P_{th} in Eq. 4.19 gives an upper bound for the estimate of the the maximum conversion efficiency available for a pure CW pump,

$$\max\{G_i^{sp}\}|_{\text{CW pump}} = \sinh^2(21\gamma A_{eff}/g_B(0)) \qquad (4.35)$$

which is independent of the fiber length. With $g_B(0)/A_{eff} = 1.5(\text{Wm})^{-1}$ and $\gamma = 10(\text{Wm})^{-1}$,

$$\max\{G_i^{sp}\}|_{\text{CW pump}} \cong -17\text{dB}. \qquad (4.36)$$

Similarly,

$$\max\{G_i^{pc}\}|_{\text{CW pump}} \cong \max\{G_i^{fc}\}|_{\text{CW pump}} \cong -13\text{dB}. \qquad (4.37)$$

These rather low conversion efficiencies lead to high noise figures as shown in Fig. 4.8 so that an increase of the maximum input pump power is required.

To increase the SBS threshold, the growth of the backwards propagating Stokes wave discussed in section 2.2.4 must be prevented. An obvious way is to directly decouple single pieces of the HNLF by optical isolators [94]. This approach is limited by the unavoidable insertion loss of the isolators that makes the scheme ineffective when using many pieces for a large increase of the threshold. Since the SBS gain spectrum is very narrow, another option is to spectrally distribute the Stokes power such that only a small part falls within the SBS gain spectrum. This can be done by either changing the Brillouin frequency shift Ω_B along the fiber or by spectrally broadening the pump wave.

The Brillouin frequency shift Ω_B can be changed by varying the core diameter [95] or the doping concentration [96, 97], by applying strain [98, 99, 87] or by applying a temperature distribution [100, 101]. The threshold increase provided by all these techniques is moderate (in the order of 5 dB) and typically

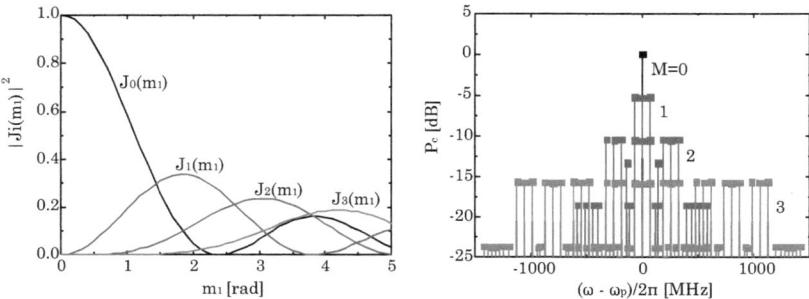

Figure 4.9: a) Squared magnitude of the Bessel functions of order 0 to 3, b) Modulated pump spectrum for different numbers M of sinusoidal tones ($m_1 = m_2 = m_3 = 1.44$, $f_1 = 69$ MHz, $f_2 = 253$ MHz, $f_3 = 805$ MHz)

also variations in the zero-dispersion wavelength are introduced degrading the FOPA performance. An exception is ref. [87] where the strain was used to equalize existing fluctuations of the zero-dispersion wavelength and increase the SBS threshold at the same time. The requirement for this was a careful measurement of the spatial λ_{zd}-distribution in the fiber.

A larger increase of the SBS threshold can be achieved by spectral distribution of the pump wave. To keep the pump power time-independent, this is done by phase- or frequency modulation [102]. The method described in the following uses a phase modulation with several sinusoidal signals and properly chosen frequencies in order to keep the spectral distance of the sidebands above the SBS bandwidth [103]. A threshold increase of more than 20 dB was achieved using 5 modulation tones with frequencies up to 10 GHz [24].

When applying several sinusoidal modulation signals to a phase modulator in the pump path as shown for the single-pump FOPA in Fig. 4.1, the modulated phase of the pump wave is given by

$$\phi_{\sin}(\mathbf{m},\mathbf{f},\boldsymbol{\theta}) = \sum_{n=1}^{M} m_n \cos(2\pi f_n t + \theta_n). \qquad (4.38)$$

where M is the number of applied modulation tones, $\mathbf{m} = (m_1,...,m_M)$ is the set of phase modulation indices of the M tones, $\mathbf{f} = (f_1,...,f_M)$ is the set of frequencies of the M tones and $\boldsymbol{\theta} = (\theta_1,...,\theta_M)$ is the set of phases of the M tones. The normalized power spectrum of the pump signal with the carrier frequency ω_p that is phase modulated with a single sinusoidal tone with frequency f_1 and modulation index m_1 is given by

$$\left| \mathscr{F}\left[(e^{im_1 \cos(2\pi f_1 t)}) \right] \right|^2 = \sum_{l=-\infty}^{\infty} |J_l(m_1)|^2 \delta(\omega - (\omega_p + l 2\pi f_1)) \qquad (4.39)$$

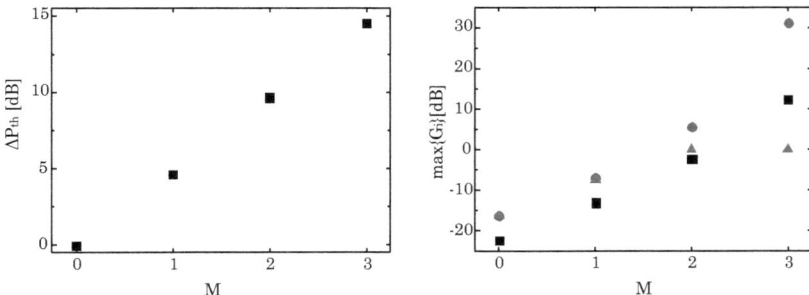

Figure 4.10: a) Increase of Brillouin threshold as a function of the number of sinusoidal tones M, b) Maximum conversion efficiency for SP (squares), PC (circles) and FC (triangles) as a function of the number of sinusoidal tones M ($m_1 = m_2 = m_3 = 1.44$, $f_1 = 69$ MHz, $f_2 = 253$ MHz, $f_3 = 805$ MHz). The used parameters are given in the text.

Here, $\mathscr{F}[\ ]$ denotes the Fourier transform, δ denotes a Dirac delta function and J_l is the l-th order Bessel function that determines the lth side band power. Fig. 4.9a shows the 0th to the 3rd order Bessel functions as a function of m_1. For $m_1 = 1.44$ the signal power is equally distributed on the carrier and the two first sidebands. At this modulation index, the SBS threshold increases by a factor 3, i.e. by nearly 5 dB if the modulation frequency is chosen more than twice the SBS bandwidth, $f_1 > 2\Delta\nu_B$. Adding a second sinusoidal with a frequency $f_2 \cong 3f_1$ and $m_2 = 1.44$ will lead to $3 \times 3 = 9$ equal sidebands. Generally, the application of M sinusoidal tones with well separated frequencies, $f_{l+1} > 3f_l$ and $f_1 \cong 2\Delta\nu_B$, will lead to 3^M spectral components with equal powers and spectral distance larger than twice the Brillouin bandwidth. Using Eq. 4.38, the spectra

$$P_c(\omega) = \left|\mathscr{F}\left[(e^{i\phi_{\sin}(t)})\right]\right|^2 \qquad (4.40)$$

are shown in Fig. 4.9b for a different number of modulation tones M. The increase of the Brillouin threshold as a function of the number of modulation tones is shown in Fig. 4.10a. It is obtained by

$$\Delta P_{\text{th}} = P_p - \max\left\{\left|\mathscr{F}\left[\sqrt{P_p}\, e^{i\phi_{\sin}(t)}\right]\right|^2 * g_B(\omega)/g_B(0)\right\} \qquad (4.41)$$

where $*$ denotes a convolution. Thus, to calculate ΔP_{th}, first the convolution of the power spectrum of the sinusoidally phase modulated pump signal with the normalized Lorentzian SBS gain spectrum given in Eq. 2.58 is calculated. Then, the difference between the total pump power and peak power of the convoluted spectrum is taken which equals the increase in Brillouin threshold. For the simulation, $\Delta\nu_B = 40$ MHz was used. The simulated values are

Table 4.1: Additional phase distortions introduced by FOPAs

Phase distortion	Affects ...	Type	Reason
Transfer of the pump-phase modulation	Idler wave	Deterministic	Pump-phase modulation for SBS suppression
Transfer of the pump laser phase noise	Idler wave	Stochastic	Phase noise from the pump laser diode
Pump-induced nonlinear phase noise	Amplified signal and idler wave	Stochastic	XPM due to amplitude noise on the pump wave
Signal-induced nonlinear phase noise	Amplified signal and idler wave	Stochastic	SPM and XPM due to amplitude noise on the input signal wave

close to the theoretical value of $10\log_{10}(3^M)$. In Fig. 4.10b, the corresponding maximum conversion efficiency is shown for the three FWM processes. The threshold power (here, in contrast to Eqs. 4.36 and 4.37, defined as the input CW power of which the reflected power reaches 1 percent in order to get more realistic values) was calculated to P_{th} = 8.8 dBm using Eq. 2.54 and the parameters L = 1km, $\Delta\nu_B = 40 MHz$, $g_B(0)/A_{eff} = 1.5 (Wm)^{-1}$ and $\gamma = 10 (Wkm)^{-1}$. $P_{\text{th,SBS}} + \Delta P_{\text{th}}$ was then inserted into the Eqs. 4.19, 4.23 and 4.27. The graph shows that the use of two modulation tone leads to conversion efficiencies of about 0 dB. Higher conversion efficiencies need more tones.

4.1.5 Additional Phase Distortions

In a previous section, the degradation of the output signal quality by complex Gaussian noise was discussed using the noise figure. However, FOPAs introduce additional deterministic and (non - Gaussian distributed) stochastic phase distortions that are not included in the noise figure. They are listed in table 4.1. One source of these distortions are imperfections of the pump signal(s). First, the pump signal has to be phase modulated in order to suppress SBS. However, this pump-phase modulation is transferred to the converted signal due to the FWM process distorting any data phase modulation. Second, the pump signal has a non-zero line width, i.e. the pump signal exhibits laser phase noise from the pump laser source as discussed in section 3.1.2. Similar to the transfer of the pump-phase modulation, this laser phase noise is also transferred to the converted signal due to the FWM distorting any data phase modulation. Third, the pump signal also exhibits some amplitude noise, either the relative-intensity noise (RIN) of the pump laser source or amplified spontaneous emission (ASE) noise from the amplification by EDFAs which is also shortly discussed in section 3.1.2. This amplitude noise translates into phase noise on the amplified and the converted signal due to XPM which will be referred to as pump-induced nonlinear phase noise. A fourth source of phase distortions are amplitude distortions on the input signal itself. They translate into phase distortions of the amplified and the converted signal and will be referred to as signal-induced phase noise in the following.

In this subsection, these four types of phase distortions will be quantified for the three different FWM processes. A detailed discussion on their impact on the signal quality of phase modulated signals in terms of BER then follows in the subsequent sections.

It should be noted that FOPAs also introduce deterministic and (non - Gaussian) stochastic amplitude distortions. Because their impact on phase-modulated signals is limited, they will be only shortly discussed in the subsequent sections in comparison to the phase distortions.

Single-Pump Process

For the single-pump process, the output phase of the converted signal can be derived from the approximate analytic solution given in Appendix H. For perfect phase matching, the phase shift due to the conversion process is given by Eq. H.24,

$$\vartheta_i^{sp} = \frac{\pi}{2} + 2\phi_p - \frac{\Delta B_{sp}}{2}L + \gamma P_p L. \tag{4.42}$$

Here, ϕ_p is the input phase of the pump, $\frac{\Delta B_{sp}}{2}L$ is the phase shift due to the chromatic dispersion and $\gamma P_p L$ is the phase shift due to XPM by the pump wave. The input phase of the pump,

$$\phi_p = \phi_{sin}(\mathbf{m}_p, \mathbf{f}_p, \boldsymbol{\theta}_p) + \phi_l(\Delta v_p), \tag{4.43}$$

comprises a sinusoidal pump-phase modulation ϕ_{sin} as defined in Eq. 4.38 used for SBS suppression and a laser phase noise contribution, ϕ_l, describing the pump laser phase noise due to the non-zero pump laser line width Δv_p as discussed in section 3.1.2. Thus, the phase shift due to the conversion process can be written as

$$\vartheta_i^{sp} = \frac{\pi}{2} - \frac{\Delta B_{sp}}{2}L + \underbrace{2\phi_{sin}(\mathbf{m}_p, \mathbf{f}_p, \boldsymbol{\theta}_p)}_{\phi_{ppm}^{sp}} + \underbrace{2\phi_l(\Delta v_p)}_{\phi_{lpn}^{sp}} + \underbrace{\gamma P_p L}_{\phi_{xpm}^{sp}} + \phi_{spm}^{sp}. \tag{4.44}$$

Here, ϕ^{sp}_{ppm} represents the phase distortion due to the transfer of the pump-phase modulation to the idler, ϕ^{sp}_{lpn} accounts for the laser phase noise transferred from the pump and ϕ^{sp}_{xpm} denotes the phase shift due to XPM by the pump wave. The last term ϕ^{sp}_{spm} takes into account the phase distortion due to SPM/XPM of the input signal and the generated idler itself. It was introduced phenomenologically because signal and idler SPM/XPM was neglected in the approximate solution in Appendix H. In a similar way, the phase shift of the amplified output signal is given by Eq. H.23,

$$\vartheta^{sp}_s = -\frac{\Delta B_{sp}}{2}L + \underbrace{\gamma P_p L}_{\phi^{sp}_{xpm}} + \phi^{sp}_{spm}. \qquad (4.45)$$

As the pump phase is not transferred to the amplified signal, no phase distortions due to the pump-phase modulation or the pump laser phase noise occur.

Phase Conjugation Process

For perfect phase matching, the phase shift of the converted signal due to the conversion process is given by Eq. H.49,

$$\vartheta^{pc}_i = \frac{\pi}{2} + \phi_{p1} + \phi_{p2} - \frac{\Delta B_{pc}}{2}L + \frac{3}{2}\gamma(P_{p1}+P_{p2})L \qquad (4.46)$$

Here, ϕ_{p1} and ϕ_{p2} are the input phases of the two pumps, $\Delta B_{pc}L/2$ is the phase shift due to the chromatic dispersion and $\frac{3}{2}\gamma(P_{p1}+P_{p2})L$ is the phase shift due to XPM by the pump waves. As for the single-pump process, the input phases of the pumps,

$$\phi_{p1} = \phi_{sin}(\mathbf{m}_{p1},\mathbf{f}_{p1},\boldsymbol{\theta}_{p1}) + \phi_l(\Delta\nu_{p1}) \qquad (4.47)$$

$$\phi_{p2} = \phi_{sin}(\mathbf{m}_{p2},\mathbf{f}_{p2},\boldsymbol{\theta}_{p2}) + \phi_l(\Delta\nu_{p2}), \qquad (4.48)$$

comprise the sinusoidal modulations $\phi_{sin}(\mathbf{m}_{p1},\mathbf{f}_{p1},\boldsymbol{\theta}_{p1})$ and $\phi_{sin}(\mathbf{m}_{p2},\mathbf{f}_{p2},\boldsymbol{\theta}_{p2})$, both given by Eq. 4.38. The second contributions, $\phi_l(\Delta\nu_{p1})$ and $\phi_l(\Delta\nu_{p2})$, describe the pump laser phase noise due to the non-zero pump laser line widths $\Delta\nu_{p1}$ and $\Delta\nu_{p2}$ and are given by Eq. 3.6. Thus, the phase shift of the idler can be written as

$$\vartheta^{pc}_i = \frac{\pi}{2} - \frac{\Delta B_{pc}}{2}L + \underbrace{\phi_{sin}(\mathbf{m}_{p1},\mathbf{f}_{p1},\boldsymbol{\theta}_{p1}) + \phi_{sin}(\mathbf{m}_{p2},\mathbf{f}_{p2},\boldsymbol{\theta}_{p2})}_{\phi^{pc}_{ppm}}$$

$$+ \underbrace{\phi_l(\Delta\nu_{p1}) + \phi_l(\Delta\nu_{p2})}_{\phi^{pc}_{lpn}} + \underbrace{\frac{3}{2}\gamma(P_{p1}+P_{p2})L}_{\phi^{pc}_{xpm}} + \phi^{pc}_{spm}. \qquad (4.49)$$

The meaning of the individual terms is the same as in the single-pump case. The phase shift of the amplified output signal is given by Eq. H.48,

$$\vartheta_s^{pc} = -\frac{\Delta B_{pc}}{2}L + \underbrace{\frac{3}{2}\gamma(P_{p1}+P_{p2})L}_{\phi_{xpm}^{pc}} + \phi_{spm}^{pc}. \qquad (4.50)$$

Similarly, no phase distortions due to the pump-phase modulation or the pump laser phase noise occur.

Frequency Conversion Process

For perfect phase matching, the phase shift of the converted signal due to the conversion process can be written with Eq. H.67 as

$$\vartheta_i^{fc} = \phi_{p1} - \phi_{p2} - \frac{\Delta B_{fc}}{2}L + \frac{5}{2}\gamma P_{p1}L + \frac{3}{2}\gamma P_{p2})L. \qquad (4.51)$$

As for the phase-conjugation process, ϕ_{p1} and ϕ_{p2} are the input phases of the two pumps given by Eqs. 4.47, $\frac{\Delta B_{fc}}{2}L$ is the phase shift due to the chromatic dispersion and $\frac{5}{2}\gamma P_{p1} + \frac{3}{2}\gamma P_{p2})$ is the phase shift due to XPM by the pump waves. Thus, the phase shift of the idler can be written as

$$\vartheta_i^{fc} = -\frac{\Delta B_{fc}}{2}L + \underbrace{\phi_{sin}(\mathbf{m}_{p1},\mathbf{f}_{p1},\boldsymbol{\theta}_{p1}) - \phi_{sin}(\mathbf{m}_{p2},\mathbf{f}_{p2},\boldsymbol{\theta}_{p2})}_{\phi_{ppm}^{fc}}$$

$$+ \underbrace{\phi_l(\Delta v_{p1}) - \phi_l(\Delta v_{p2})}_{\phi_{lpn}^{fc}} + \underbrace{\frac{5}{2}\gamma P_{p1}L + \frac{3}{2}\gamma P_{p2}L}_{\phi_{xpm}^{fc}} + \phi_{spm}^{fc}. \qquad (4.52)$$

The meaning of the individual terms is the same as in the previous cases. The phase shift of the attenuated output signal can be derived from Eq. 4.53,

$$\vartheta_s^{fc} = \frac{\Delta B_{fc}}{2}L + \underbrace{\frac{5}{2}\gamma P_{p1} + \frac{3}{2}\gamma P_{p2}L}_{\phi_{xpm}^{fc}} + \phi_{spm}^{fc}. \qquad (4.53)$$

Similarly, no phase distortions due to the pump-phase modulation or the pump laser phase noise occur.

Cascaded Amplification and Wavelength Conversion

If an input signal is amplified (or attenuated) N_c times by similar FOPAs (i.e., which use the same FWM process but are not necessarily identical in terms of field gain and phase shift [4]), the ratio of the output signal of the last stage,

---

[4] Of course, also cascaded amplification (or attenuation) using different FWM processes is possible, which will not be treated here for simplicity.

$A_{s,N_c}(L)$, to the input signal of the first stage, $A_{s,1}(0)$, is given by

$$\frac{A_{s,N_c}(L)}{A_{s,1}(0)} = \prod_{l=1}^{N_c} \mathscr{G}_{s,l} \exp\left(i\vartheta_{s,l}\right) \quad (4.54)$$

under the assumption of perfect phase matching where $\mathscr{G}_{s,l}$ is the field gain of the l-th amplifier and $\vartheta_{s,l}$ is the phase shift due to the l-th amplifier. The formula is valid for all three FWM processes. The accumulated phase shift of the output signal is therefore given by

$$\Theta_{s,N_c} = \sum_{l=1}^{N_c} \vartheta_{s,l}, \quad (4.55)$$

i.e., the individual phase contributions simply add up. For cascaded wavelength conversions, the ratio of the output signal of the last stage to the input signal of the first stage is given by

$$\frac{A_{i,N_c}(L)}{A_{s,1}(0)} = \prod_{l=1}^{N_c} \mathscr{G}_{i,l} \exp\left(i(\pm 1)^{N_c-l}\vartheta_{s,l}\right) \quad (4.56)$$

under the assumption of perfect phase matching where $\mathscr{G}_{i,l}$ is the field conversion efficiency of the l-th wavelength converter and $\vartheta_{s,l}$ is the phase shift due to the l-th wavelength converter. If the minus sign is chosen in the factor $(\pm 1)^{N_c-l}$, the formula is valid to describe cascaded wavelength conversion by the single-pump and the phase-conjugation process. Its appearance is a consequence of the fact that both processes produce phase conjugated idlers [5]. The plus sign is valid for the (not phase conjugating) frequency-conversion process. The accumulated phase shift of the output signal is given by

$$\Theta_{i,N_c} = \sum_{l=1}^{N_c} (\pm 1)^{N_c-l}\vartheta_{i,l}. \quad (4.57)$$

which is again a sum of the individual contributions.

4.2 Laser Phase Noise

In this section, the phase distortion of the wavelength converted signal due to the laser phase noise discussed in section 3.1.2 of the pump wave(s) is examined. It does not affect the amplified signal and is characterized by the terms ϕ_{lpn}^{sp}, ϕ_{lpn}^{pc} and ϕ_{lpn}^{fc} for the three different FWM processes in Eqs. 4.44, 4.49 and 4.52, respectively.

---

[5] Note that the output signal of N_c cascaded wavelength conversions using the single-pump or the phase-conjugation process is not phase conjugated if N_c is an even number, while it is phase conjugated if N_c is an odd number.

In direct detection systems, the laser phase noise is given by the transmitter laser and leads to a phase error in the interferometric comparison of two subsequent symbols with a noise variance given in Eq. 3.5 and $\tau = T_s = 1/R_s$. In coherent detection systems, the laser phase noise of the received signal is the sum of the phase noise contributions of the transmitter and the receiver laser, i.e. the linewidth is doubled if the same laser type is used, and causes a random walk of the phase reference that has to be compensated for using the carrier phase estimation algorithm described in section 3.2.2. Although both mechanisms are quite different, they can be both simply characterized by defining a required laser linewidth that gives a signal (O)SNR penalty [6] of 1 dB at a BER = 10^{-4} in comparison to the case without laser phase noise. This linewidth (per laser for the coherently detected formats) normalized to the bit rate is given in table 4.2 for the different modulation formats [6, p. 165, p. 193]. The results have been obtained by Monte-Carlo simulations. For the coherently detected formats, the feed forward m-th power scheme described in Sec. 3.2.2 was applied as the phase estimation algorithm using optimized average block lengths. Two general tendencies can be identified: First, the directly detected formats are more tolerant to laser phase noise than the coherently detected formats. Second, the higher-order formats are less tolerant due to their smaller Euclidian symbol distance and due to their lower symbol rate at the same bit rate because, after Eq. 3.5, a higher $\tau = T_s$ increases the variance of the phase difference between two consecutive symbols.

4.2.1 Single-Pump Configuration

In the single-pump configuration, the laser phase noise of the pump is given by $\phi_p = \phi_l(\Delta v_p)$ as defined in Eq. 3.6. The resulting phase distortion of the wavelength converted signal is given by Eq. 4.44, $\phi_{lpn}^{sp} = 2\phi_p$. Adding also the laser phase noise of transmitter laser, $\phi_t \phi_l(\Delta v_t)$, which is independent of the pump laser phase noise, and using Eq. 3.5, the linewidth of the idler after a single conversion is given by

$$\Delta v_i = \Delta v_t + 2\Delta v_p, \qquad (4.58)$$

i.e. the pump laser linewidth is added twice to the transmitter laser linewidth. Assuming generally equal linewidths Δv_l for transmitter, receiver and pump laser, the joined linewidth for directly detected formats is

$$\Delta v_{l,DD} = (2N_c + 1)\Delta v_l \qquad (4.59)$$

---

[6]Since the SNR penalty describes the relative increase of the required SNR to reach a certain BER, it is equal to the OSNR penalty as defined in section 1.1.

Modulation format	$\Delta v_l/$ (Bit rate)
DBPSK	10^{-2}
DQPSK	10^{-3}
4-PSK	10^{-4}
8-PSK	10^{-5}
Square 16-QAM	10^{-6}

Table 4.2: (Average) laser linewidth normalized to the bit rate required for an (O)SNR penalty < 1dB at a BER of 10^{-4} for different modulation formats [6, p. 165, p. 193]. The results were obtained by Monte-Carlo simulations. For the directly detected formats (DBPSK, DQPSK), the required transmitter laser linewidth is given. For the coherently detected formats (4-PSK, 8-PSK, square 16-QAM), the average laser linewidth (averaged over transmitter and receiver laser) is given and the feed forward m-th power algorithm with optimized average block length described in Sec. 3.2.2 was applied as the phase estimation algorithm in the simulations.

and for coherently detected formats

$$\Delta v_{l,CD} = (2N_c + 2)\Delta v_l \qquad (4.60)$$

where N_c is the number of conversions. Fig. 4.11a shows the required laser linewidth for different bit rates for a single conversion and for the back-to-back case. While the requirements for DBPSK and DQPSK are quite relaxed also at low bit rates, they are very high at low bit rates for 8-PSK and 16-QAM relaxing only slightly for higher bit rates. While the requirements are already high for the back-to-back case, the wavelength conversion enlarges this present problem. This is even more pronounced for cascaded wavelength conversions as shown in Fig. 4.11b for a bit rate of 100 Gb/s.

4.2.2 Dual-Pump Configuration

For the dual-pump configuration, the linewidths of pump 1 and pump 2 are given Δv_{p1} and Δv_{p2}. The phase distortion of the wavelength converted signal is given by Eqs. 4.49 and 4.52, $\phi_{lpn} = \phi_l(\Delta v_{p1}) \pm \phi_l(\Delta v_{p2})$, where the plus sign refers to the phase-conjugation process and the minus sign to the frequency-conversion process. Adding also the laser phase noise of the transmitter laser and assuming independent laser phase contributions, the idler linewidth using Eq. 3.5 is given by

$$\Delta v_i = \Delta v_t + \Delta v_{p1} + \Delta v_{p2}, \qquad (4.61)$$

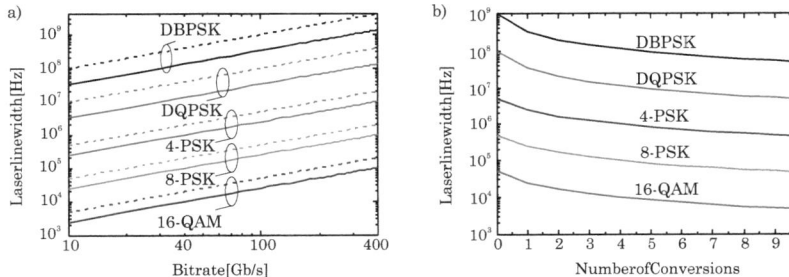

Figure 4.11: a) Required laser linewidth (averaged over the transmitter, receiver and the pump laser(s)) for 1 dB signal (O)SNR penalty @ BER = 10^{-4} compared to the case without laser phase noise for different bit rates and modulation formats (dashed line - back-to-back, solid line - single conversion), b) same as a) but for a bit rate of 100 Gb/s and different numbers of conversions. The calculations used Eqs. 4.59 and 4.60 and the values given in Table 4.2.

Thus, the laser linewidth requirements for the dual-pump configuration are identical to those for the single-pump configuration given in Fig. 4.11. However, by using two pumps with correlated phase noise, the frequency-conversion process gives the interesting opportunity to create an idler without increased laser phase noise because the phase noise contributions of the two pumps are subtracted in the idler phase. This can be realized by using a single laser source and creating two pumps by amplitude modulation [104] or by extracting the pumps from a mode-locked laser [105].

4.3 Impact of the Pump-Phase Modulation

In this section, the impact of the deterministic phase distortion due to the pump-phase modulation on various phase-modulation formats will be studied. Similarly to the laser phase noise of the pump(s), it does not affect the amplified signal but only the wavelength converted signal and is characterized by the terms ϕ_{ppm}^{sp}, ϕ_{ppm}^{pc} and ϕ_{ppm}^{fc} for the three different FWM processes in Eqs. 4.44, 4.49 and 4.52, respectively.

4.3.1 Single-Pump Configuration with Direct Detection

Single Conversion

The setup for the characterization of the single-pump FOPA is shown in Fig.

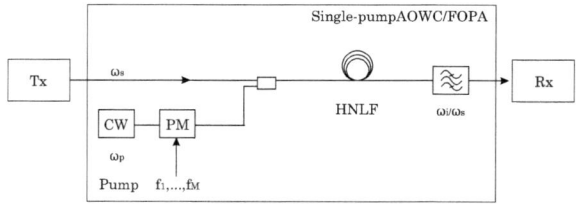

Figure 4.12: Single-pump FOPA setup to determine the impact of the pump-phase modulation

4.12. For the single-pump process, the phase distortion of the wavelength converted signal due to the pump-phase modulation is given by Eq. 4.44,

$$\phi_{ppm}^{sp} = 2\phi_{sin}(\mathbf{m}_p, \mathbf{f}_p, 0) = 2\sum_{n=1}^{M} m_n \cos(2\pi f_n t) \qquad (4.62)$$

where $\theta_p \equiv 0$ was assumed without loss of generality. D(Q)PSK signals are detected by comparing the phases of two subsequent symbols in a delay interferometer with a delay of one symbol period $T_s = 1/R_s$. The differential phase distortion, i.e. the phase difference between two consecutive symbols, is given by [42]

$$\Delta\phi_{ppm}^{sp} = \phi_{ppm}^{sp}(t+T_s) - \phi_{ppm}^{sp}(t) \qquad (4.63)$$

$$\approx \frac{d\phi_{ppm}^{sp}}{dt} T_s \qquad (4.64)$$

$$= -4\pi \sum_{n=1}^{M} m_n f_n T_s \sin(2\pi f_n t) \qquad (4.65)$$

where the approximation holds as long as $f_n \ll R_s$. Its maximum is then given by

$$\max\{\Delta\phi_{ppm}^{sp}\} = 4\pi \sum_{n=1}^{M} m_n f_n / R_s \approx 4\pi m_M f_M / R_s. \qquad (4.66)$$

The maximum differential phase distortion is proportional to the modulation frequencies and the modulation index and inversely proportional to the symbol rate. It is dominated by the highest modulation frequency f_M for which $\max\{\Delta\phi_{ppm}^{sp}\}$ is shown as a function of in Fig. 4.13a. The corresponding (O)SNR penalty due to the pump-phase modulation at a BER of 10^{-4} is shown in Fig. 4.13b. Since the maximum differential phase distortion is dominated by the highest modulation frequency, it was analogously assumed for the BER calculation that

$$\Delta\phi_{ppm}^{sp} \approx -4\pi m_M f_M T_s \sin(2\pi f_M t). \qquad (4.67)$$

The BER was calculated by interpreting $\Delta\phi_{ppm}^{sp}$ as a time-dependent interferometer phase error. Then, Eq. 3.27 can be used by averaging over one

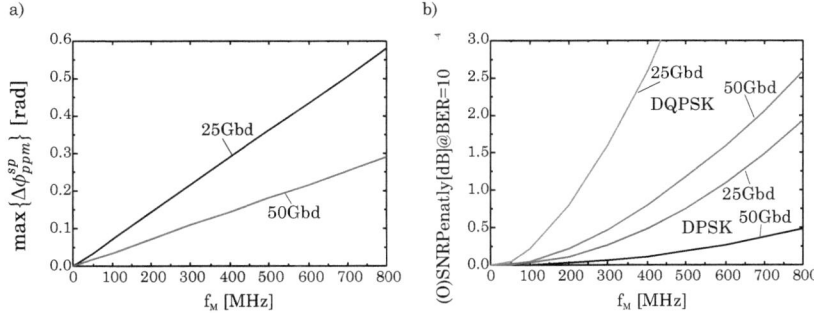

Figure 4.13: a) Maximum differential phase distortion, b) corresponding (O)SNR penalty for D(Q)PSK with different symbol rates, $m_M = 1.44$ pump-phase modulation period $T_M = 1/f_M$, [42]

$$\text{BER} = \frac{1}{T_M} \int_0^{T_M} \text{BER}(\text{SNR}_s, \Delta\phi_{ppm}^{sp}) dt. \qquad (4.68)$$

The (O)SNR penalties are quickly growing in all cases with the modulation frequency. For DPSK, they are still moderate but for DQPSK, they are larger due to the higher sensitivity to phase distortions. This behavior was validated by experimental results [106, 42]. The penalty is decreasing for higher symbol rates due to the decreasing differential phase distortion. As an example, one can assume that the (O)SNR penalty shall be kept below 1 dB. Then, the maximal modulation frequency should be chosen below 250 MHz and 500 MHz for 25 GBd and 50 GBd DQPSK, respectively, as can be seen in Fig. 4.13b. This restriction limits the maximum available conversion efficiency. Fig. 4.10 shows that a conversion efficiency of about 0 dB can be obtained using a maximal modulation frequency of about 250 MHz, while a higher conversion efficiency needs 800 MHz. Thus, in the example, the conversion efficiency is limited to 0 dB if the (O)SNR penalty shall be kept below 1 dB.

Multiple Conversions

In the following, the accumulation of phase distortions during N_c cascaded wavelength conversions shall be discussed. For simplicity, it is again assumed that the phase distortion is dominated by the highest modulation frequency. Furthermore, the contributions of the different wavelength converters shall be identical except of a random, in the interval $[0, 2\pi]$ uniformly distributed relative phase θ_l. Then, the phase distortion due to the l-th single-pump wave-

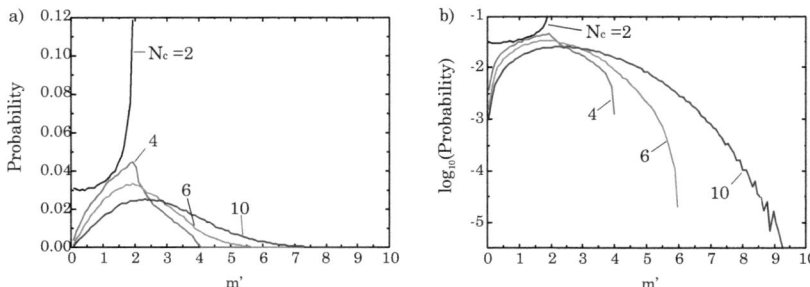

Figure 4.14: a) PDF of the multiplier m' for different number of conversions N_c, b) same as a) but in logarithmic style

length converter, $\phi_{ppm,l}^{sp}$, is given by

$$\phi_{ppm,l}^{sp} = 2m_M \cos(2\pi f_M t + \theta_l). \tag{4.69}$$

now taking into account explicitly that the sinusoidal tones of the different stages are not added in phase, but with random phases θ_l. The accumulated contribution Θ_{ppm,N_c}^{sp} can be calculated using Eq. 4.57,

$$\Theta_{ppm,N_c}^{sp} = \sum_{l=1}^{N_c} (\pm 1)^{N_c-l} \, \phi_{ppm,l}^{sp}. \tag{4.70}$$

Because the random phases θ_l are uniformly distributed in the interval $[0, 2\pi]$, the factor $(-1)^{N_c-l}$ can be omitted and the accumulated contribution can then be written as [107, 37]

$$\begin{aligned} \Theta_{ppm,N_c}^{sp} &= 2m_M \sum_{l=1}^{N_c} \cos(2\pi f_M t + \theta_l) \\ &= 2m_M \Re \left\{ \sum_{l=1}^{N_c} \exp(2\pi i f_M t + i\theta_l) \right\} \\ &= 2m_M \, \cos(2\pi f_M t + \xi) \underbrace{\left| \sum_{l=1}^{N_c} \exp(i\theta_l) \right|}_{m'}. \end{aligned} \tag{4.71}$$

The last factor acts as a multiplier for the modulation index m_M and will be called m', and ξ represents the sum's complex phase. Since, according to Eq. 4.63, the phase distortion due to the pump-phase modulation is proportional to the modulation index, $m' > 1$ will lead to a further signal degradation. The PDF of m' is shown in Fig. 4.14 in both linear and logarithmic style for different numbers of wavelength conversions N_c. The best case is given by $m' = 0$ and the worst case is given by $m' = N_c$. While for $N_c = 2$ the worst case is very likely to occur, its probability quickly decreases for higher values of N_c and

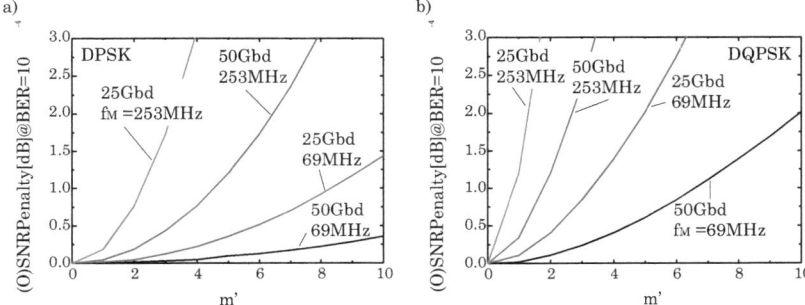

Figure 4.15: a) DPSK (O)SNR Penalty as a function of m' for different baud rates and modulation frequencies, b) same as a) but for DQPSK ($m_M = 1.44$). The calculations used Eqs. 4.68 and 4.67 with $m_M \equiv m' m_M$.

the mean value of m' grows much slower than N_c. However, the worst case probability does not drop to zero as seen from Fig. 4.14b. The corresponding (O)SNR penalties are shown in Fig. 4.15a and b for DPSK and DQPSK, respectively. They were calculated using Eqs. 4.68 and 4.67 with $m_M \equiv m' m_M$. The maximum modulation frequency and the symbol rate have been varied. The penalties grow very fast with m' (which corresponds to the number of conversions N_c in the worst case), in particular for $f_M = 253$ MHz (corresponding to a conversion efficiency of about 0 dB). For this maximum modulation frequency, $N_c \cong 5$ for DPSK and $N_c \cong 3$ for DQPSK seem possible. Assuming that m' changes slowly with time one can also define maximum outage probabilities to relax the SNR requirements [37].

4.3.2 Optical Compensation Using the Dual-Pump Configuration

In dual-pump AOWCs, the phase distortion due to the pump-phase modulation is given by Eqs. 4.49 and 4.52, $\phi_{ppm}^{pc/fc} = \phi_{sin}(\mathbf{m}_{p1}, \mathbf{f}_{p1}, \boldsymbol{\theta}_{p1}) \pm \phi_{sin}(\mathbf{m}_{p2}, \mathbf{f}_{p2}, \boldsymbol{\theta}_{p2})$ (with plus and minus referring to the phase-conjugation and the frequency-conversion process, respectively) and can be in principle avoided since the phases of the two pumps can be adjusted individually [108, 94, 43, 109]. The requirement for this is counterphasing of the pumps for the phase-conjugation process,

$$\phi_{sin}(\mathbf{m}_{p1}, \mathbf{f}_{p1}, \boldsymbol{\theta}_{p1}) = -\phi_{sin}(\mathbf{m}_{p2}, \mathbf{f}_{p2}, \boldsymbol{\theta}_{p2}), \qquad (4.72)$$

and cophasing for the frequency-conversion process,

$$\phi_{sin}(\mathbf{m}_{p1}, \mathbf{f}_{p1}, \boldsymbol{\theta}_{p1}) = \phi_{sin}(\mathbf{m}_{p2}, \mathbf{f}_{p2}, \boldsymbol{\theta}_{p2}). \qquad (4.73)$$

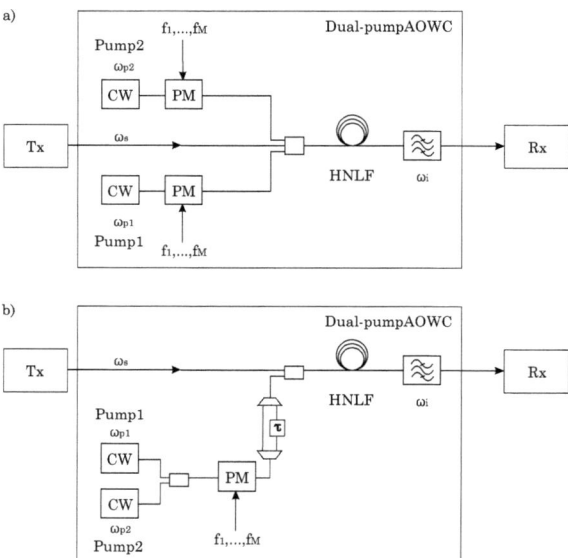

Figure 4.16: Dual-pump FOPA setup a) with two separate phase modulators for the two pumps, b) with a single phase modulator for the two pumps (and optionally two wavelength selective couplers and a delay line)

Two possible configurations are shown in Fig. 4.16. In Fig. 4.16a, two separate phase modulators are used for the two pumps allowing for independent modulation. A second configuration is shown in Fig. 4.16b [110, 43]. Here, both pumps are modulated in the same modulator resulting in cophasing. To realize counterphasing, the two pumps have to be temporally delayed by half the period of the phase modulation. In the modulation scheme discussed in section 4.1.4, this corresponds to a delay of $1/(2 \times 23$ MHz).

Ideal co- or counterphasing is difficult to realize in practice. A mismatch of the modulation indices of the two pumps can occur due to differences in two phase modulators in Fig. 4.16a or due to the spectral dependency of the modulation response of the phase modulator in Fig. 4.16b. A mismatch in the temporal alignment of the phase modulations can occur due to a mismatch in the two pump paths in Fig. 4.16a or in the delay used in Fig. 4.16b. Therefore, the tolerances will be discussed in the following. Only the highest modulation frequency will be taken into account, since it dominates the phase distortion as discussed for the single-pump configuration. Then, for the phase-conjugation process, the non-ideal pump phases are given by Eq. 4.38

$$\phi_{\sin}(\mathbf{m}_{p1},\mathbf{f}_{p1},\boldsymbol{\theta}_{p1}) \cong -m_M \cos(2\pi f_M t) \qquad (4.74)$$

$$\phi_{\sin}(\mathbf{m}_{p2},\mathbf{f}_{p2},\boldsymbol{\theta}_{p2}) \cong m_M(1+\Delta m)\cos(2\pi f_M t + \Delta\vartheta) \qquad (4.75)$$

The modulation-index mismatch Δm represents possibly different modulation indices and the pump-phase mismatch $\Delta\vartheta$ accounts for a non-ideal phase shift between the pumps. The pump-phase contribution to the idler phase distortion is given by

$$\begin{aligned}\phi_{ppm}^{pc} &= \phi_{\sin}(\mathbf{m}_{p1},\mathbf{f}_{p1},\boldsymbol{\theta}_{p1}) + \phi_{\sin}(\mathbf{m}_{p2},\mathbf{f}_{p2},\boldsymbol{\theta}_{p2}) \\ &= -2m_M \sin(\Delta\vartheta/2)\sin(2\pi f_M t + \Delta\vartheta/2) \\ &\quad + m_M \Delta m \cos(2\pi f_M t + \Delta\vartheta) \qquad (4.76)\\ &\approx m_M[-\Delta\vartheta \sin(2\pi f_M t) + \Delta m \cos(2\pi f_M t)], \qquad (4.77)\end{aligned}$$

where the approximation holds for $\Delta\vartheta \ll 1$. It is easy to see that ϕ_{ppm}^{pc} only vanishes for ideal counterphasing, $\Delta m \equiv \Delta\vartheta \equiv 0$. In all other cases, the idler is still modulated with f_M and an effective modulation index that increases linearly with the modulation-index and the pump-phase mismatch. Similar to Eq. 4.66, the resulting maximal phase distortion for a converted signal is given by

$$\max\{\phi_{ppm}^{pc}\} \approx 2\pi m_{M,\mathrm{eff}} \frac{f_M}{R_s} \qquad (4.78)$$

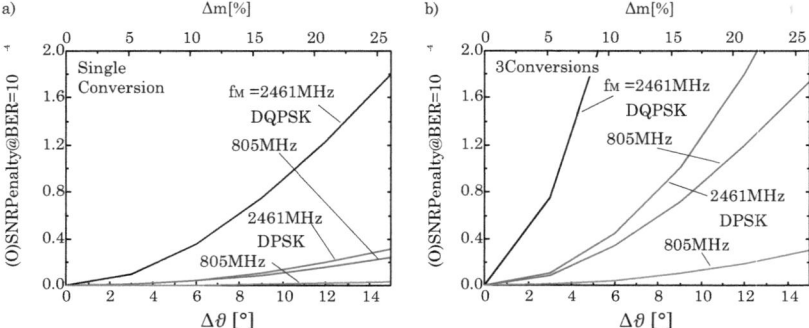

Figure 4.17: a) Single conversion (O)SNR penalty for BER = 10^{-4} as a function of the pump-phase mismatch $\Delta\vartheta$ and modulation index mismatch Δm for 25 GBd DPSK and DQPSK, b) same as a) but for 3 conversions ($m_M \to m_M m'$, $m' = 3$). For both graphs, $m_M = 1.44$ was used.

with the effective modulation index $m_{M,\text{eff}} = m_M \Delta m$ in the case of pure modulation-index mismatch and $m_{M,\text{eff}} = m_M \Delta\vartheta$ in the case of pure pump-phase mismatch. It is important to note that (4.76) - (4.78), although derived for the phase-conjugation process and counterphasing, also apply for a conversion using the frequency-conversion process where cophasing is used.

Similarly to section 4.3.1, the (O)SNR penalty can be calculated by using Eqs. 4.68 and 4.76. In Fig. 4.17a, the results for pure pump-phase mismatch (corresponding to lower x-axis) and for pure modulation-index mismatch (corresponding to upper x-axis) are plotted for a single conversion of 25 GBd DPSK and DQPSK signals and for two different modulation frequencies. For DQPSK, the use of a high maximum modulation frequency results in strict tolerances for the pump-phase and the modulation-index mismatch. This behavior was validated in experiments [111]. For a lower f_M, as well as for DPSK, the tolerances are more relaxed. For multiple conversions, as shown in Fig. 4.17b for m' = 3, the tolerances decrease quickly.

If the phase mismatch occurred due to a delay $\Delta\tau$ before the HNLF, e.g. due to length differences in the two pump paths shown in Fig. 4.16a or due to an error in the delay shown in Fig. 4.16b, it is given by

$$\Delta\vartheta = 2\pi f_M \Delta\tau \qquad (4.79)$$

and the maximal phase distortion,

$$\max\{\phi_{ppm}\} \approx 4\pi^2 m_M \Delta\tau \frac{f_M^2}{R_s}, \qquad (4.80)$$

increases quadratically with the maximum modulation frequency f_M. Fig.

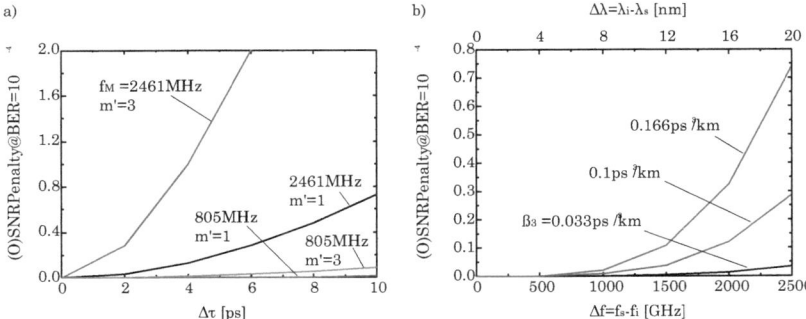

Figure 4.18: a) (O)SNR penalty for BER = 10^{-4} for 25 GBd DQPSK as a function of $\Delta\tau$ for different maximum modulation frequencies and different m', b) (O)SNR penalty for BER = 10^{-4} for FC-based wavelength conversion of 25 GBd DQPSK as a function of the conversion bandwidth for different values of the third-order dispersion coefficient β_3, a maximum modulation frequency of f_M = 2461 MHz and a fiber length L = 1 km. For both graphs, m_M = 1.44 was used.

4.18a shows the (O)SNR penalty against the delay error for 25 GBd DQPSK signals and for two different maximum modulation frequencies. For f_M = 2461 MHz, a 1-dB penalty occurs for a 10-ps delay error. For cascaded wavelength conversions, the tolerances decrease. Pump-phase mismatch due to a delay error can also occur due to walk-off for the pumps in dispersive elements. E.g., for 60 nm pump spacing, a group delay difference of 1 ps occurs in a patch cord comprising a standard single-mode fiber with length of 1 m.

Also the walk-off between the pumps in the HNLF itself due to the group-velocity dispersion should be considered. Using Eq. 2.19, the walk-off is given by [43]

$$\Delta\tau_0(z) = \left(\frac{d\beta(\omega_{p1})}{d\omega} - \frac{d\beta(\omega_{p2})}{d\omega}\right)z \qquad (4.81)$$

$$= \frac{\beta_3}{2} z \left[(\omega_{p1} - \omega_{zd})^2 - (\omega_{p2} - \omega_{zd})^2\right] \qquad (4.82)$$

where the Taylor expansion of the propagation constant $\beta(\omega)$ was performed around the zero-dispersion wavelength, $\omega_0 = \omega_{zd}$ and β_4 was neglected. Eq. 4.81 shows that the walk-off between the pumps is negligible for the phase-conjugation process because the pumps are placed nearly symmetrically around the zero-dispersion wavelength, $\omega_{p1} - \omega_{zd} \cong \omega_{zd} - \omega_{p2}$. However, the walk-off is unavoidable for full tunable operation in the Frequency Conversion process. In this case,

$$\Delta\tau_0(z) = \beta_3\, z\, \Delta\omega(\Delta\omega/2 - \Delta\omega_s) \tag{4.83}$$

with $\Delta\omega = \omega_s - \omega_i$ the conversion bandwidth and $\Delta\omega_s = \omega_s - \omega_{zd}$. Thereby, the frequency relation Eq. 4.9 for the Frequency Conversion process was used. Inserting the walk-off in Eq. 4.79, the phase mismatch $\Delta\vartheta$ varies along the fiber,

$$\Delta\vartheta(z) = 2\pi f_M \Delta\tau_0(z), \tag{4.84}$$

as well as the pump-phase contribution to the idler phase which is given analogously to Eq. (4.76) by

$$\phi_{ppm}^{fc}(z) = -m_M \Delta\vartheta(z)\sin(2\pi f_M t). \tag{4.85}$$

Here, $\Delta m = 0$ was assumed. To calculate the pump-phase contribution to the idler phase at the fiber output, Eq. H.51 is used. The growth of the Frequency Conversion idler A_i is proportional to [43]

$$dA_i \sim \exp\left[i(\phi_{p2} - \phi_{p1})\right] dz \tag{4.86}$$

$$\sim \exp\left[-i\phi_{ppm}^{fc}(z)\right] dz. \tag{4.87}$$

when considering only the pump-phase modulation. Then, integration yields

$$\begin{aligned}A_i(L) &\sim \int_0^L \exp\left[-i\phi_{ppm}^{fc}(z)\right] dz \\ &\sim L\, \exp\left[-i\frac{\phi_{ppm}^{fc}(L)}{2}\right] \frac{\sin\left[\phi_{ppm}^{fc}(L)/2\right]}{\left[\phi_{ppm}^{fc}(L)/2\right]}\end{aligned} \tag{4.88}$$

$$\approx \exp\left[-i\frac{\phi_{ppm}^{fc}(L)}{2}\right] \tag{4.89}$$

$$= \exp\left[i\frac{m_M \Delta\vartheta(L)}{2}\sin(2\pi f_M t)\right] \tag{4.90}$$

if $\Delta\vartheta(z)(L) \ll 1$. In comparison to Eq. 4.76, the idler phase distortion is only half. Thus, a delay due to walk-off inside the HNLF results in only half of the idler phase distortion compared to the same delay before the HNLF. Fig. 4.18 shows the (O)SNR penalty as a function of the conversion bandwidth for 25 GBd DQPSK and for different third-order dispersion coefficients β_3, respectively. The walk-off was maximum by setting $\Delta\omega_s = 0$. It can be seen that the walk-off induced idler phase modulation is not critical for standard HNLFs with $\beta_3 = 0.033\,\text{ps}^3/\text{km}$.

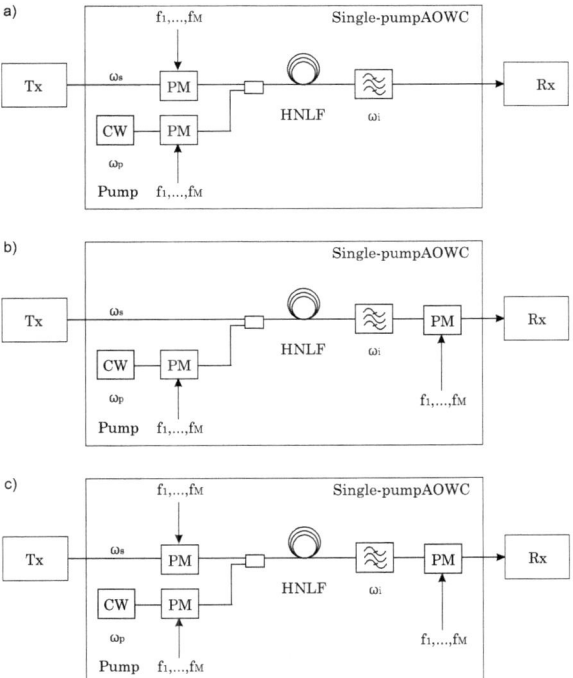

Figure 4.19: Wavelength conversion of D(Q)PSK with the single-pump FOPA and an additional phase modulator(s) placed in the signal path: a) Precompensation, b) Postcompensation, c) Pre- and postcompensation of the phase distortions due to the pump-phase modulation

4.3.3 Optical Compensation Using the Single-Pump Configuration

Another option to avoid the contribution of the pump-phase modulation to the idler phase distortion is to use the single-pump setup with one or two additional phase modulator(s) placed in the signal path. Three possible configurations are shown in Fig. 4.19. To achieve a zero idler phase distortion, the contribution of the pump-phase modulation given by Eq. 4.42,

$$\phi_{ppm}^{sp} = 2\phi_{\sin}(\mathbf{m}_p, \mathbf{f}_p, \boldsymbol{\theta}_p), \qquad (4.91)$$

has to be compensated by the additional phase modulators. In the precompensation scheme shown in Fig. 4.19a, the signal has to be modulated by ϕ_{ppm}^{sp} [112, 113]. In the postcompensation scheme shown in Fig. 4.19b, the idler has to be modulated by $-\phi_{ppm}^{sp}$. Finally, in the pre-/postcompensation scheme [114], the signal is modulated by $\phi_{ppm}^{sp}/2$ while the idler is modulated by $-\phi_{ppm}^{sp}/2$.

In practice, there are some differences between these schemes. Because phase modulators are lossy devices, the postcompensation scheme is advantageous in terms of the noise figure at high conversion efficiencies. Regarding the maximum tolerable electrical power at the phase modulator, the pre-/postcompensation scheme has an advantage because it reduces the electrical power needed to drive to the phase modulators by a factor 4.

Ideal compensation is difficult as discussed above for the dual-pump scheme. When assuming similar non-ideal modulation indices and temporal alignment characterized by Δm and $\Delta\theta$, respectively, the residual idler phase modulation can be calculated,

$$\phi_{ppm}^{sp,PM} \approx 2m_M \left[-\Delta\vartheta \sin(2\pi f_M t) + \Delta m \cos(2\pi f_M t) \right] \qquad (4.92)$$

which applies to all three discussed schemes. The residual phase distortion of the idler is now greater by a factor 2 compared to residual idler phase modulation for the dual-pump FOPA given in Eq. (4.76). Fig. 4.20 compares the corresponding (O)SNR penalties for different f_M for the single-pump FOPA with additional phase modulator(s) and for the dual-pump FOPA. For the same f_M, the tolerances for the single-pump FOPA are more strict. Furthermore, for a similar conversion efficiency as the dual-pump FOPA, the single-pump FOPA has to use a higher f_M as shown in Fig. 4.10b. Then, the advantage in tolerance of the dual-pump FOPA is even more evident.

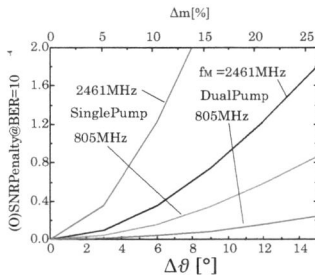

Figure 4.20: a) Single conversion (O)SNR penalty for BER = 10^{-4} as a function of the pump-phase and modulation index mismatch for 25 GBd DPSK and DQPSK, b) same as a) but for 3 conversions (m' = 3)

4.3.4 Single-Pump and Dual-Pump Configurations with Coherent Detection

In the coherent receiver, the idler is detected by mixing with a local oscillator (LO) laser and sampled at the sampling instants $t_n = n/R_s$ where n is an integer and R_s is the symbol rate. The decision for a symbol is based on the comparison of the symbol phase with the reference carrier phase. Since in an intradyne receiver as described in section 3.2.2 the LO laser is not phase-locked to the idler, the carrier phase is digitally recovered by a feed-forward carrier phase estimation (CPE) algorithm as described in section 3.2.2. As shown in Fig. 3.3b, for m-ary PSK formats, the CPE is done by first raising the detected samples to the m-th power in order to remove the data phase information. Then, a running average over N_{av} symbols is performed to minimize (zero mean) additive white Gaussian (AWG) noise. Finally, the phase is taken with an unwrapping arg-operation and divided by m. Using Eq. 3.12 and only taking into account a phase distortion due to the pump-phase modulation similar to Eq. 4.62 (i.e. ignoring the Gaussian noise and any other phase distortion including the LO laser phase noise and the carrier frequency offset), the normalized complex phasor of the wavelength converted signal after coherent reception and sampling at sampling instants $t_k = kT_s = k/R_s$ is given by [7]

$$s(k) = \frac{\tilde{X}_k}{|\tilde{X}_k|} = e^{i\phi_{ppm}(k)} = \exp\left(m_{M,i}\cos(2\pi k f_M/R_s)\right) \quad (4.93)$$

---

[7]For simplicity, the sampling rate was set equal to the symbol rate. Furthermore, it is assumed that the separation of the polarization modes (if needed), the equalization of transmission impairments and the timing recovery was ideally performed.

where only the modulation tone with the highest frequency f_M was taken into account. $m_{M,i}$ is the modulation index of the converted signal. It is e.g. given by $m_{M,i} = 2m_M$ for the single-pump FOPA as given in Eq. 4.62, by $m_{M,i} = m_M \Delta\theta$ for the dual-pump FOPA with pure phase mismatch and by $m_{M,i} = m_M \Delta m$ for the dual-pump FOPA with pure modulation-index mismatch, both given by Eq. 4.78. The PSK phase information was already omitted because it disappears anyway after raising $s(k)$ to the m-th power,

$$s(k)^m = \exp\left(m m_{M,i} \cos(2\pi k f_M/R_s)\right). \quad (4.94)$$

Most importantly, the contribution of the pump-phase modulation is enlarged by the factor m. It is instructive to discuss three different regimes:

- $m m_{M,i} \gg 1$ In this regime, the signal raised to the m-th power is strongly phase modulated and its bandwidth is given by the Carson rule,

$$B = 2 f_M (m m_{M,i} + 2) \quad (4.95)$$

If $B > R_s$ the Nyquist criterion is violated because the signal bandwidth is higher than the sample rate. In time domain, this means that the maximum phase shift between two samples exceeds the range $[-\pi,+\pi]$ so that phase ambiguities occur. Then a proper carrier phase estimation is not possible.

- $m m_{M,i} \approx 1$ Here, the signal phase is widespread on the unity circle and averaging over a significant part of the modulation frequency period, $N_{av} \approx R_s/f_M$ will result in a strong decrease in the phasor magnitude. This will cause the CPE to fail. Thus, $N_{av} \ll R_s/f_M$ must hold. However, this degrades the AWGN suppression leading to a high probability of cycle slips. Thus, the CPE is working unstable.

- $m m_{M,i} \ll 1$ In this regime, the CPE is possible also for high N_{av} leading to a good AWG noise suppression and stable operation.

From this classification, it is clear that the proper coherent detection of any PSK signal converted by the single-pump FOPA without optical compensation by additional phase modulators is hardly possible. From the typical value of $m_M = 1.44$ rad follows $m_{M,i} = 2.88$ rad so that the first or the second regime apply here. Another option to solve this issue is additional electronic signal processing, as described in the next section.

Thus, the following discussion in this section applies only to wavelength converters with a reduced contribution of the pump-phase modulation to the idler phase, i.e. the dual-pump FOPA with co- or counterphasing of the pumps or

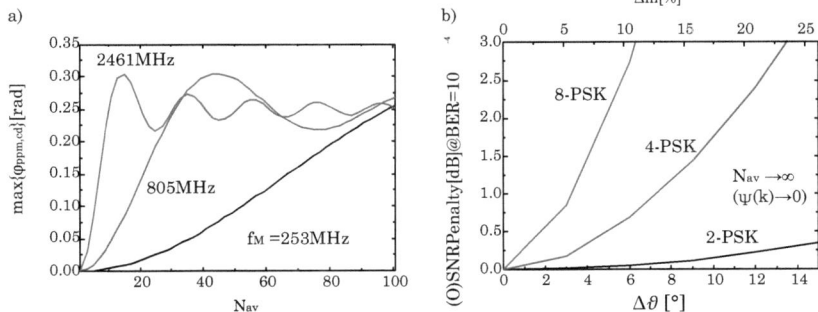

Figure 4.21: a) Maximum phase error as a function of the number of averaged samples in the CPE algorithm for different maximum modulation frequencies (25 GBd, 4-PSK, $m_{M,i} = 0.25$) b) (O)SNR penalty for BER = 10^{-4} after the dual-pump converter as a function of the pump-phase and modulation index mismatch for coherently detected 25 GBd 2-PSK, 4-PSK and 8-PSK ($N_{av} \to \infty \equiv \psi(k) \to 0$, $m_M = 1.44$)

the single-pump FOPA with additional phase modulator(s), as described in the previous sections.

For these AOWCs, the third regime applies, for which the phase distortion after the CPE and its impact on the BER can be calculated analytically. As shown in Appendix J, the carrier phase ψ_k recovered by the m-th power algorithm is given by Eq. J.9,

$$\psi(k) = \frac{\sin\left(\frac{\pi N_{av} f_M}{R_s}\right)}{N_{av} \sin\left(\frac{\pi f_M}{R_s}\right)} \phi_{ppm}(k). \qquad (4.96)$$

Ideally, the recovered carrier phase should incorporate the whole phase distortion due to the pump-phase modulation, i.e. it would be identical to the phase distortion ϕ_{ppm}, in order to leave an undistorted signal phase after the phase recovery. Thus, the remaining phase distortion after the phase recovery is given by

$$\phi_{ppm,cd}(k) = \phi_{ppm}(k) - \psi(k) = \left(1 - \frac{\sin\left(\frac{\pi N_{av} f_M}{R_s}\right)}{N_{av} \sin\left(\frac{\pi f_M}{R_s}\right)}\right) m_{M,i} \cos(2\pi k f_M / R_s). \qquad (4.97)$$

The maximum phase distortion $\max\{\phi_{ppm,cd}\}$ for 25 GBd 4-PSK is shown in Fig. 4.21a as a function of the number of averaged symbols N_{av} ($m_{M,i} = 0.25$ rad). For $N_{av} = 1$, no remaining phase distortion is left because no averaging takes place. However, as discussed above, N_{av} must be chosen large in order to cancel out (zero-mean) AWG noise (which was not included in the calculation within this section). As N_{av} is increased, also the (zero-mean) phase modulation averages out leading to an increase in the remaining phase distortion eventually saturating at the original phase distortion before phase recovery, ϕ_{ppm}, after some oscillations. Although one can imagine a compromise since a high N_{av} is needed for good AWG noise suppression and a low N_{av} leads to lower phase distortions, one would usually choose a high N_{av} for stable operation which results in phase distortion approximately equal to the original phase distortion of the idler ($N_{av} \to \infty \equiv \psi(k) \to 0$). The resulting BER can be calculated by inserting $\phi_{PPM,cd}(k) = \phi_{ppm}(k)$ into Eq. 3.29 and averaging of the modulation period $T_M = 1/f_M$,

$$\text{BER} = \frac{1}{T_M} \int_0^{T_M} \text{BER}(\text{SNR}_s, \phi_{PPM,cd}) dt. \quad (4.98)$$

The (O)SNR penalty for a signal converted by a co- or counterphased dual-pump FOPA (ϕ_{ppm} is given by Eq. 4.77 in this case) is shown in Fig. 4.21b as a function of pump-phase mismatch $\Delta\theta$ (lower x-axis) and the modulation index mismatch Δm (upper x-axis). The penalties due to pump-phase and modulation-index mismatch are independent of the maximum modulation frequency and larger than those for the directly detected formats (compare to DQPSK in Fig. 4.17a). Thus, similar to the tolerance against laser phase noise, the tolerance against the phase distortion due to the pump-phase modulation is larger for directly detected formats than for coherently detected formats. Still, coherent detection gives the unique opportunity for electronic compensation of the phase distortions, as shown in the next section.

4.3.5 Compensation Using Electronic Signal Processing

In the previous section, three different regimes were defined characterized by the magnitude of the product $mm_{M,i}$. Only if $mm_{M,i} \ll 1$, the CPE algorithm is working stable. For $mm_{M,i} \approx 1$, the CPE algorithm is unstable since it is impossible to optimize the suppression of AWG noise and the phase distortions simultaneously by varying the averaging window. However, coherent detection gives the unique opportunity to circumvent the problem by using electronic signal processing to compensate the phase distortion before the CPE algorithm [115, 45].

The flow chart of a compensation algorithm is shown in Fig. 4.22. The samples s(k) after coherent detection, sampling and electronic equalization [8] represent the complex idler symbols at a sample rate of 1 sample per symbol.

---

[8] Including separation of the polarization modes (if needed), equalization of transmission impairments and timing recovery

For a higher A/D-sample rate, the idler field must be down-sampled. After a carrier frequency offset correction, the algorithm extracts the phase of the complex idler symbols raised to the m-th power and uses it to estimate the parameters of the phase distortion consisting of M sinusoidals (frequency f_n, phase θ_n and modulation index $m_{n,i}$) [116] on a block of N_i incoming data samples by Fast Fourier Transform (FFT). An example is shown in Fig. 4.23. Here, the phase distortion consists of two sinusoidals with frequencies of 69 MHz and 253 MHz. As can be seen, the procedure has to be iterated two times since, after the first iteration, the higher frequency component of the phase distortion is not completely suppressed. The reason for this behavior are cycle slips occuring during the phase extraction due to the presence of Gaussian noise. The cycle slips become manifest in Fig. 4.23a in the discontinuities of the phase before the compensation and lead to a too low estimate of $m_{n,i}$. The algorithm converges since the probability for cycle slips decreases while the residual phase modulation decreases. The iteration is stopped if the estimate of all residual modulation indices is less than 0.1 rad [9]. This condition ensures stable behavior also for very small values of $m_{n,i}$ for which the estimates are more vulnerable to the Gaussian noise. To increase the accuracy of the parameter extraction beyond the frequency grid of the FFT, quadratic interpolation is used for the estimation of the modulation indices $m_{n,i}$ and the frequencies f_n as well as linear interpolation for the estimation of the phase θ_n is used. As an example, for a sampling rate of 25 GHz [10], $N_i = 2^{17}$ and padding with 3 N_i zeros, the FFT has a resolution of about $\Delta f_{\text{FFT}} = 50$ kHz. Using quadratic interpolation, the frequency accuracy can be increased to below 2 kHz. To estimate the frequency f_n, the maximum of the FFT power spectrum $|(\text{FFT}(y))|^2$ (y defined as in Fig. 4.22) is detected within the search window $[0.9\ \tilde{f}_n, 1.1\ \tilde{f}_n]$ where $\tilde{f}_n$ represents a rough estimate of f_n. The maximum may be given by the value pair $(f_b, \underline{b})$. The adjacent values are given by $(f_b - \Delta f_{\text{FFT}}, \underline{a})$ and $(f_b + \Delta f_{\text{FFT}}, \underline{c})$. Using the derivation given in Appendix K following the quadratically interpolated FFT method (QIFFT) [116] [11], the frequency estimate is given by

$$\text{est}\{f_n\} = f_b + \frac{\Delta f_{\text{FFT}}}{2} \frac{\underline{a} - \underline{c}}{\underline{a} - 2\underline{b} + \underline{c}}. \tag{4.99}$$

---

[9] Or if the estimates of $m_{n,i}$ start to increase (i.e. if the algorithm does not converge monotonically) which is not shown in Fig. 4.22

[10] That corresponds to a symbol rate of 25 GBd because 1 sample per symbol is required as mentioned above.

[11] This method is used to estimate parameters of sinusoidals in the audio technology.

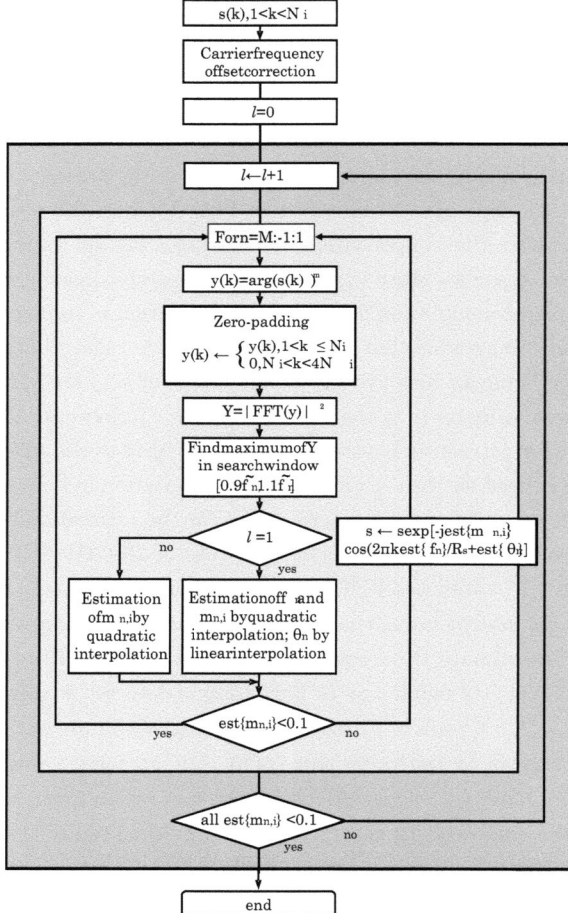

Figure 4.22: Flow chart of the compensation algorithm

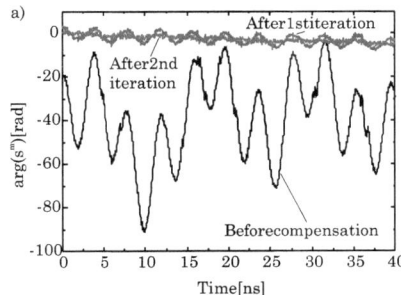 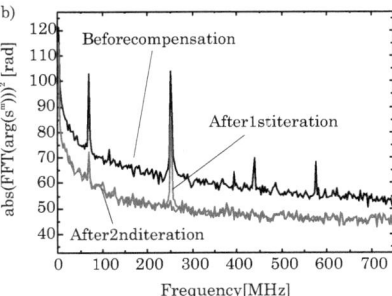

Figure 4.23: Extracted phase arg(s^m) in a) time domain and b) frequency domain (25 GBd 8-PSK, f_1 = 69 MHz, f_2 = 253 MHz, $m_{1,i} = m_{2,i}$ = 2.8 rad, 19 dB SNR, 500 kHz joint laser linewidth, no carrier frequency offset)

Again with Appendix K, the estimate of the modulation index is given by [12]

$$\text{est}\{m_{n,i}\} = \underline{b} + \frac{\text{est}\{f_n\} - f_b}{4\pi m \Delta f_{\text{FFT}}}(\underline{a} - \underline{c}). \tag{4.100}$$

The phase θ_n is estimated by linear interpolation. If θ_b, θ_a and θ_c correspond to the values of the FFT phase spectrum arg(FFT(y)) (y defined as in Fig. 4.22) at f_b, $f_b - \Delta f_{\text{FFT}}$ and $f_b + \Delta f_{\text{FFT}}$, respectively, the interpolated phase is given by

$$\text{est}\{\theta_n\} = \theta_b + (\theta_c - \theta_a)\frac{\text{est}\{f_n\} - f_b}{2\Delta f_{\text{FFT}}}. \tag{4.101}$$

Because any carrier frequency offset given by $\Delta\omega_{\text{LO}}$ in Eq. 3.10 distorts the estimation procedure it has to be compensated first. The flow chart of this algorithm as well as an example is shown in Fig. 4.24. The frequency offset correction uses the fact that a sinusoidal phase modulation produces a symmetric power spectrum. The power spectrum of the idler symbols raised to the m-th power,

$$Y_1(f) = |\text{FFT}(s^m)|^2, \tag{4.102}$$

is mirrored with respect to the zero frequency, $Y_2(f) = Y_1(-f)$, and a cross-correlation is performed between Y_1 and Y_2,

$$Y_c(\Delta f_{\text{CCF}}) = \int_{-\infty}^{\infty} Y_1(f + \Delta f_{\text{CCF}})Y_2(f)df \tag{4.103}$$

The frequency value of the cross-correlation peak max(Y_c) corresponds to twice the frequency offset estimate $\Delta\omega_{\text{LO}}$.

Numerical simulations have been performed to test the tolerance of the algorithm against laser phase noise and Gaussian noise. In Fig. 4.25, the mismatch of the parameter estimation for 25 GBd 8-PSK and a phase distortion

---

[12] Note that this is dependent on the definition of the discrete Fourier transform

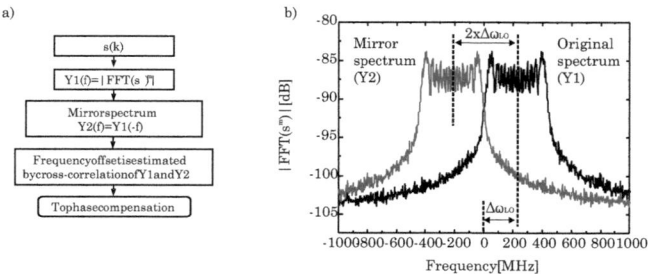

Figure 4.24: Flow chart of the frequency offset compensation algorithm

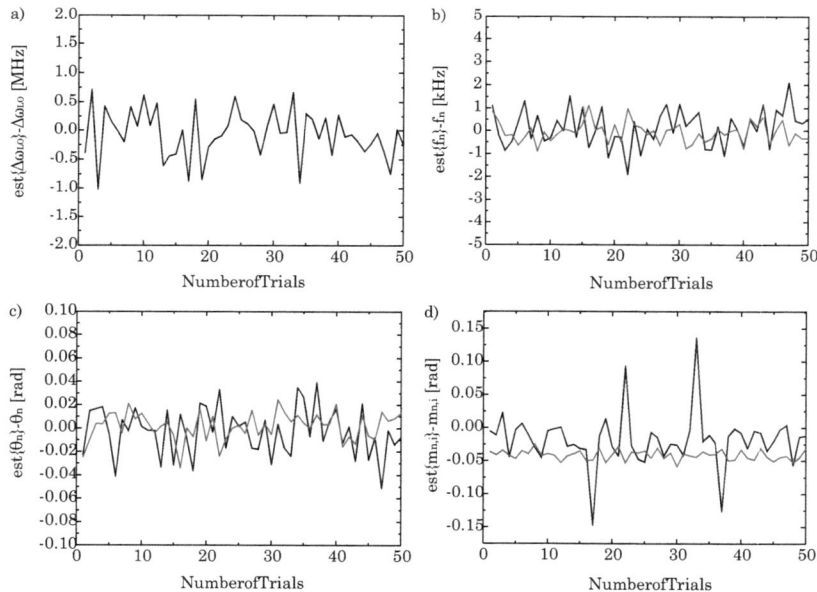

Figure 4.25: Parameter estimate mismatch while using the compensation algorithm shown in Fig 4.22 for 25Gbd 8-PSK distorted by two sinusoidal modulation tones (black - 69 MHz, red - 253 MHz) and 50 numerical simulations. The used parameters are given in the text.

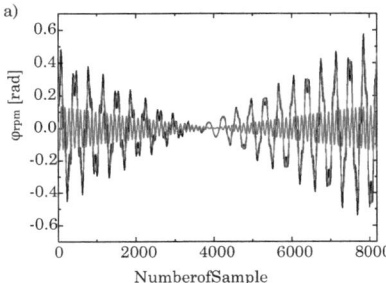 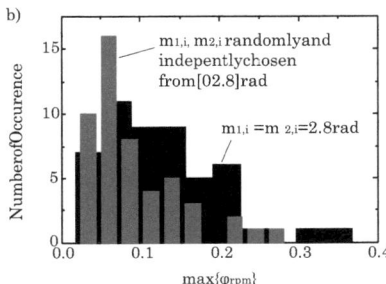

Figure 4.26: a) Residual phase distortion of the idler symbols after the compensation algorithm for 25 GBd 8-PSK, 2 modulation tones with 69 and 253 MHz and 2^{13} samples (black - total, red - 69 MHz, green - 253 MHz), b) Histogram of the maximum residual phase modulation index for two modulation tones (69 MHz and 253 MHz), $N_i = 2^{18}$ and 50 numerical simulations

consisting of two modulation tones (f_1 = 69 MHz, f_2 = 253 MHz, $m_{1,i} = m_{2,i}$ = 2.8 rad) is shown for 50 numerical simulations with 2^{18} test samples. The carrier frequency offset was chosen randomly (< 500 MHz) as well as the phases θ_1 and θ_2 of the two sinusoidal tones. In each trial, the test signal was distorted by different random Gaussian noise realizations at an SNR of 16 dB. The joined laser linewidth (test signal plus local oscillator) was set to 500 kHz. The estimate mismatch is always small indicating the stable operation of the algorithm. The necessary number of iterations used by the algorithm was 4 for all trials.

The upper limit on the amount of phase modulation that can be compensated is given by the Nyquist condition Eq. 4.95 as discussed in the previous section. This sets a fundamental limitation on the product of PSK order, maximum modulation frequency and its modulation index. As an example, for 25 GBd 8-PSK and a modulation index of $m_{M,i}$ = 2.8 rad, the maximum modulation frequency which can be compensated is given by about f_M = 500 MHz while for 25 GBd QPSK and $m_{M,i}$ = 2.8 rad, it is given by f_M = 950 MHz. A close look on the resulting residual phase distortion φ_{rpm} of the idler symbols after the compensation algorithm (and before the CPE algorithm) is shown in Fig. 4.26a. There is still a small sinusoidal modulation with a modulation index varying over the block length. This is due to the small estimate mismatch in modulation frequency which results in a beat pattern. The total residual phase modulation is dominated by the lower frequency component because the estimation of its parameters is more difficult due to the lower number of sine periods within the data block (compare to Fig. 4.25). His-

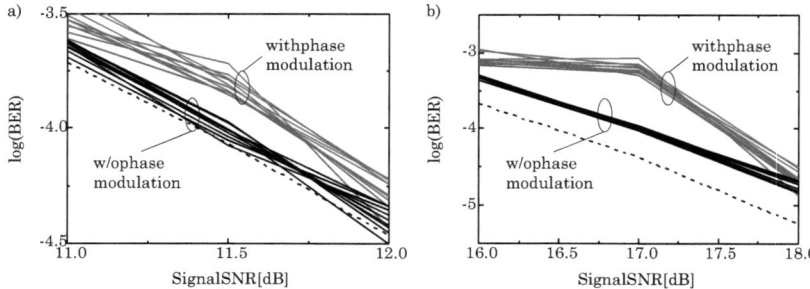

Figure 4.27: a) BER for 25 GBd 4-PSK with 3 modulation tones (69 MHz, 253 MHz, 805 MHz) with $m_{1,i} = m_{2,i} = m_{3,i} = 2.8$ rad, b) BER for 25 GBd 8-PSK with 2 modulation tones (69 MHz, 253 MHz) with $m_{1,i} = m_{2,i} = 2.8$ rad ($N_i = 2^{19}$, joined laser linewidth 100kHz, carrier frequency offset randomly chosen from [0 500MHz], $N_{av} = 49$)

tograms of the maximum residual phase modulation index max$\{\varphi_{rpm}\}$ for two different cases and 50 numerical simulations are shown in Fig. 4.26b. The black columns correspond to the same modulation indices $m_{1,i} = m_{2,i} = 2.8$ in each simulation. As can be estimated from Fig. 4.25, max$\{\varphi_{rpm}\}$ varies from simulation to simulation due to the random Gaussian and laser phase noise realizations. However, the variation range is small with a mean of about 0.1 rad. The higher values stem from a estimate mismatch of the 69 MHz modulation tone. A further narrowing of the distribution may be achieved by reusing the information about the estimated parameter from former data blocks because the pump-phase modulation will change slowly in comparison to the bit time scale. The red columns correspond to a case where the modulation indices of the two modulation tones were randomly and independently chosen from the interval [0 2.8] assuming a uniform distribution. As discussed in section 4.3.1, such random modulation indices can occur after cascading several (dual-pump) AOWCs. The resulting distribution of the maximum residual phase modulation index is comparable to the first case with constant modulation indices (black columns) indicating that the estimation process works for any modulation index. Thus, the compensation algorithm can also be used to equalize modulation index variations of the phase modulation in the cascaded operation of AOWCs.

Fig. 4.27 shows BER curves calculated with the Monte Carlo method using $N_i = 2^{19}$ samples. Fig. 4.27a shows the BER for 25Gbd 4-PSK with 3 modulation tones and 10 numerical simulations. The variations are due to the random Gaussian and laser phase noise realizations and nearly equal with and without the phase modulation. This shows that the variation of the residual phase distortion has minor impact on the BER because the dominating variations of the 69 MHz components are effectively reduced by the CPE algorithm itself as seen from Fig. 4.21a. The penalty at BER = 10^{-4} is about 0.3 dB. Fig. 4.27b shows the BER for 25 GBd 8-PSK with 2 modulation tones and 10 numerical simulations. Here, the penalty at BER = 10^{-4} is slightly higher with 0.6 dB. These results show that the compensation algorithm enables highly efficient wavelength conversion of higher-order phase modulation signals as validated by experiments [115, 45].

4.3.6 Comparison to Impact on Amplitude Modulated Signals

The pump-phase modulation also impacts amplitude modulated signals via the gain G_s and the conversion efficiency G_i, respectively. This was investigated for OOK signals in a series of papers [117, 118, 119, 120, 121, 122, 123]. One can differentiate two different effects. First, the gain is dependent on the phase matching parameter κ and therefore on the pump frequency. Since the phase modulation is equivalent to a frequency modulation, this leads to a gain modulation at the pump-phase modulation frequencies [117, 119]. Second, any dispersive element as the HNLF itself or the pump filters does convert phase variations into amplitude variations. These pump amplitude variations cause gain variations due to the pump power dependence of the gain [124, 120]. Both effects are significant only at high gain levels and depend on the bandwidth of the pump-phase modulation. That means that only high-frequency components introduce significant distortions, e.g. when using PRBS-modulated BPSK sequences for the pump-phase modulation. In this way, the use of several sinusoidals for the pump-phase modulation is optimal to reduce the amplitude fluctuations. This is confirmed by measurements showing a minor impact on the BER of OOK signals [118, 122]. Moreover, the use of exactly counterphased pumps for the phase-conjugation based FOPA further reduces the amplitude fluctuations [119, 124]. This is the same optimum operation condition as for the reduction of the phase distortions. From these results, it is expected that for mixed amplitude and phase modulation formats like 16QAM, the phase distortion due to the pump-phase modulation will dominate over the amplitude distortion although no investigations have been performed yet.

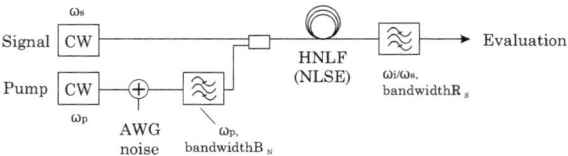

Figure 4.28: Single-pump FOPA simulation setup for characterization of pump-induced noise

4.4 Pump-Induced Noise

In this section, the phase distortions due to XPM by the pump wave(s) will be discussed. This is accounted for by the phase shift ϕ_{xpm} in the Eqs. 4.44, 4.45, 4.49, 4.50, 4.52 and 4.53. The XPM effect was explained in section 2.1.3. It leads to nonlinear phase noise on the wavelength converted idler wave as well as on the amplified signal wave if the pump wave exhibits amplitude noise [125, 126, 46, 127, 128]. The noise may be due to relative intensity noise from the laser diode or due to ASE noise from amplification. Furthermore, also nonlinear amplitude noise will be discussed in this section. Its origin is also pump amplitude noise that is transferred via the power-dependence of the gain / conversion efficiency.

4.4.1 Pump-Induced Phase Noise in the Single-Pump Configuration

For the single-pump configuration, the phase shift due to pump XPM for the amplified signal and the idler is given in Eqs. 4.44 and 4.45 as

$$\phi_{xpm}^{sp} = \gamma P_p L. \qquad (4.104)$$

In the following, it is assumed that the pump wave is distorted by noise. Then, the envelope can be rewritten as

$$A_p = <A_p> + \Delta A_p \qquad (4.105)$$

where, without loss of generality, a real-valued mean amplitude $<A_p>$ and a complex valued, zero-mean fluctuation term ΔA_p were assumed. Since the fluctuations shall be small, $|\Delta A_p| \ll <A_p>$, the pump power is approximately given by

$$P_p = |A_p|^2 = \underbrace{<A_p>^2}_{<P_p>} + \underbrace{2<A_p> \Re\{\Delta A_p\}}_{\Delta P_p} \qquad (4.106)$$

where $\Re\{\}$ denotes the real part. The average pump power is given by $<P_p> = <A_p>^2$ while the pump power fluctuations are given by $\Delta P_p = 2<A_p>\Re\{\Delta A_p\}$. Consistent with Eqs. 1.1 and 3.7, the pump signal-to-noise ratio (pump SNR) can be defined using Eq. (4.105) as

$$\text{SNR}_p = \frac{<P_p>}{2\langle\Re\{\Delta A_p\}^2\rangle} = \frac{2<P_p>^2}{<(\Delta P_p)^2>}. \tag{4.107}$$

If the pump wave is directly generated by a high power CW laser, the pump power fluctuations are dominated by the relative intensity noise (RIN) of the laser that is due to spontaneous emission of radiation into the laser mode. The pump SNR can be related to the electrically measured RIN by

$$\int_0^\infty \text{RIN}(f)df = \frac{\langle(\Delta P_p)^2\rangle}{<P_p>^2} = \frac{2}{\text{SNR}_p^{\text{RIN}}} \tag{4.108}$$

where $\langle(\Delta P_p)^2\rangle = 4<P_p>\langle\Re\{\Delta A_p\}^2\rangle$ is the pump power mean square calculated by using Eq. (4.106). Assuming a constant RIN spectrum within an electrical bandwidth $B_N/2$ and a fast decrease beyond gives a rough estimate for the pump SNR,

$$\text{SNR}_p^{\text{RIN}} \approx \frac{2}{\text{RIN} \times \min(B_N/2, R_s/2)}. \tag{4.109}$$

The term $\min(B_N/2, R_s/2)$ was introduced to emphasize that nonlinear phase noise outside the signal bandwidth is rejected by the receiver filter. In this way, the effective pump noise bandwidth corresponds to the electrical receiver filter bandwidth in maximum (which is ideally half of the symbol rate, $R_s/2$)[13]. If, as usual in laboratory experiments, the pump waves are amplified by erbium-doped fiber amplifiers (EDFAs) before entering the HNLF, there will be also an amplified spontaneous emission (ASE) noise contribution from the EDFA. If the ASE contribution is dominating, the pump SNR is given by

$$\text{SNR}_p^{\text{ASE}} = \frac{<P_p>}{\rho_{\text{ASE}} \min(B_N, R_s)}. \tag{4.110}$$

with the ASE power spectral density ρ_{ASE}. B_N is the optical bandwidth of the bandpass filter after the EDFA. There is a third noise contribution due to the quantum noise that is, however, negligible assuming high pump powers > 100 mW. Then, the overall pump SNR is given by

$$\frac{1}{\text{SNR}_p} = \frac{1}{\text{SNR}_p^{\text{RIN}}} + \frac{1}{\text{SNR}_p^{\text{ASE}}} \tag{4.111}$$

Both RIN and ASE noise contributions imply that the fluctuation term $\Re\{\Delta A_p\}$ is normally distributed. Furthermore, within the following sections, it will

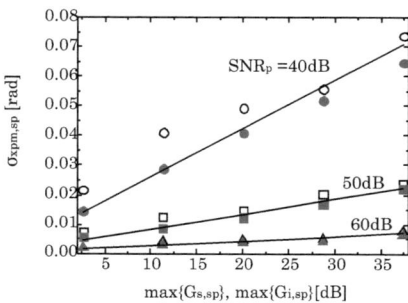

Figure 4.29: Standard deviation of the pump-induced XPM phase shift in the single-pump process for signal and idler as a function of the maximum gain and different pump SNR (solid line: theory after Eq. 4.116, open symbols: simulation results for the signal, filled symbols: simulation results for the idler. The used parameters are given in the text.

be always assumed that $R_s > B_N$ leading to symbol rate independent signal (O)SNR penalties. Using Eqs. 4.104 and 4.106, the variance of the XPM phase shift can be derived,

$$\begin{aligned}\sigma^2_{xpm,sp} &= <(\phi^{sp}_{xpm})^2> - <\phi^{sp}_{xpm}>^2 \\ &= \gamma^2 L^2 (<P_p^2> - <P_p>^2) \\ &= \gamma^2 L^2 <\Delta P_p^2> \\ &= 4\gamma^2 L^2 <P_p> \langle \Re\{\Delta A_p\}^2\rangle. \end{aligned} \quad (4.112)$$

Using Eq. 4.107, this can be rewritten to

$$\sigma^2_{xpm,sp} = 2\gamma^2 L^2 \frac{<P_p>^2}{\text{SNR}_p} \quad (4.113)$$

$$= 2\frac{<\phi^{sp}_{xpm}>^2}{\text{SNR}_p} \quad (4.114)$$

The nonlinear phase noise variance $\sigma^2_{xpm,sp}$ is proportional to the square of the mean XPM phase distortion $<\phi^{sp}_{xpm}>$ and inversely proportional to the pump signal-to-noise ratio. According to Eq. 4.19, also the FOPA gain increases with the product $\gamma P_p L$. Thus, increasing the FOPA gain means to increase the nonlinear phase noise variance for the same pump SNR. In order to obtain a lower bound on the noise variance that can be expected for a FOPA with a certain gain, a high gain / conversion efficiency and perfect phase matching is assumed. In this case, the maximum gain for the single-pump configuration

---
[13]This is true if Nyquist signaling [129, p. 545] is assumed.

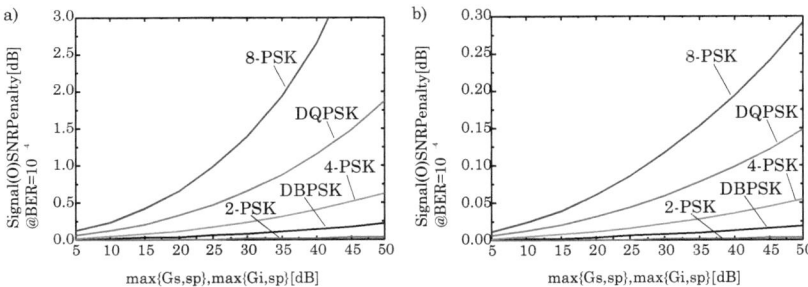

Figure 4.30: Signal (O)SNR penalty @ BER = 10^{-4} for different phase modulation formats as a function of $\max\{G^{sp}_{s,i}\}$ and a) SNR_p = 40 dB and b) SNR_p = 50 dB

given in Eq. 4.19 can be approximated by

$$\max\{G_i^{sp}\} \cong \max\{G_s^{sp}\} \cong \frac{1}{4}\exp(2\gamma <P_p> L). \tag{4.115}$$

Insertion into Eq. 4.113 gives

$$\sigma^2_{xpm,sp} \cong \frac{\ln^2(4\max\{G_s^{sp}\})}{2\,SNR_p} \tag{4.116}$$

Remarkably, the variance is fully parametrized by the maximal parametric gain and the pump SNR. Fig. 4.29 shows the standard deviation $\sigma_{xpm,sp}$ as a function of the average pump power and for different values of the pump SNR. The solid lines are calculated with Eq. 4.116 while the symbols correspond to results from numerical simulations using the NLS equation (2.39) with a CW input signal and a CW pump signal. The used parameters were L = 1 km, P_s = -30 dBm, α = 0, γ = 10 (W km)$^{-1}$, λ_{zd} = 1553 nm, $\lambda_p - \lambda_{zd}$ = 1.1 nm, β_3 = 0.033 ps^3/km and β_4 = 2.5 × 10^{-4} ps^4/km. λ_s was adjusted to the gain peak for different pump powers. The output signal was optically filtered by a 2nd order Gaussian bandpass filter with 25 GHz bandwidth before evaluating the noise variances. The analytical results show a good agreement with the simulation. Fig. 4.30 shows the (O)SNR penalty for different phase modulation formats as a function of the maximum gain /conversion efficiency. The BER was calculated by inserting the nonlinear phase noise variance given in Eq. 4.116 into Eqs. 3.32 and 3.34. Generally speaking, the impact of the nonlinear phase noise increases with the number of constellation points. Also, for the same number of constellation points, the differentially modulated DPSK formats perform worse than the corresponding PSK format. This is because for DPSK formats two noisy bits are compared. While the penalties are very small for a pump SNR of 50 dB, the higher-order formats show measurable

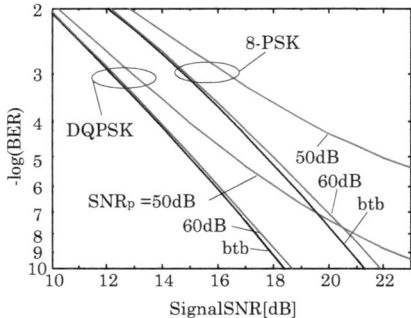

Figure 4.31: BER as a function of the signal SNR for DQPSK and 8-PSK after 10 cascaded AOWCs stages with max$\{G^{sp}_{s,i}\}$ = 40 dB in the back-to-back case and different pump SNR

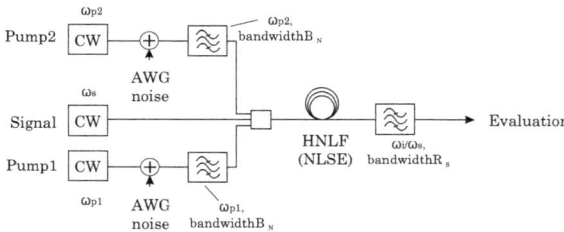

Figure 4.32: Dual-pump FOPA simulation setup for characterization of pump-induced noise

penalties for SNR$_p$ = 40 dB and high gain values. For cascaded operation, the phase shift due to pump XPM adds up as given in Eqs. 4.55 and 4.57. Since the pump noise in the different single-pump FOPAs is uncorrelated, the non-linear phase noise variances of N_c stages add up,

$$\sigma^2_{xpm,sp}\big|_{\Sigma N_c} = \sum_{l=1}^{N_c} \sigma^2_{xpm,sp}\big|_l \qquad (4.117)$$

For equal single stage variances, the resulting variance is just $N_c \times \sigma^2_{xpm,sp}$. Fig. 4.31 shows the BER for DQPSK and 8-PSK after 10 cascaded FOPA stages with $max\{G^{sp}_{s,i}\}$ = 40 dB, i.e. with gain comparable to an EDFA. To avoid high penalties, a pump SNR of 60 dB is necessary. Using Eq. 4.109, this can be translated into a RIN < -160 dB/Hz assuming a 10 GHz electrical RIN bandwidth. This high requirement on the pump wave noise level relaxes significantly at lower values for the gain /conversion efficiency.

4.4.2 Pump-Induced Phase Noise in the Dual-Pump Configuration

In the previous section, it was shown that pump-induced nonlinear phase noise is foremost a problem for FOPAs with high gains/ conversion efficiencies using high pump powers. As seen from Fig. 4.7, only moderate pump powers are needed for the frequency-conversion process because the maximal gain/ conversion efficiency for the frequency-conversion process is limited to unity. Thus, the impact of the pump-induced nonlinear phase noise is expected to be low for this process. Thus, only the phase-conjugation process will be discussed in the following.

For this process, the signal and idler phase shift due to pump XPM is given by Eqs. 4.49 and 4.50,

$$\phi_{xpm}^{pc} = \frac{3}{2}\gamma(P_{p1}+P_{p2})L. \qquad (4.118)$$

Similarly to Eq. 4.105, the pump envelopes can be defined as

$$A_{p1} = <A_{p1}> + \Delta A_{p1} \qquad (4.119)$$
$$A_{p2} = <A_{p2}> + \Delta A_{p2}. \qquad (4.120)$$

with the real-valued mean amplitudes $<A_{p1}>$ and $<A_{p2}>$ and the independent complex, zero-mean fluctuation terms ΔA_{p1} and ΔA_{p2}. When assuming $|\Delta A_{p1}| \ll <A_{p1}>$ and $|\Delta A_{p2}| \ll <A_{p2}>$, the noisy pump powers are given by

$$P_{p1} = |A_{p1}|^2 = \underbrace{<A_{p1}>^2}_{<P_{p1}>} + \underbrace{2<A_{p1}>\Re\{\Delta A_{p1}\}}_{\Delta P_{p1}} \qquad (4.121)$$

$$P_{p2} = |A_{p2}|^2 = \underbrace{<A_{p2}>^2}_{<P_{p2}>} + \underbrace{2<A_{p2}>\Re\{\Delta A_{p2}\}}_{\Delta P_{p2}} \qquad (4.122)$$

In the same manner as Eq. 4.107, the pump signal-to-noise ratios are found to be given by

$$\mathrm{SNR}_{p1} = \frac{<P_{p1}>}{2\langle\Re\{\Delta A_{p1}\}^2\rangle} \qquad (4.123)$$

$$\mathrm{SNR}_{p2} = \frac{<P_{p2}>}{2\langle\Re\{\Delta A_{p2}\}^2\rangle}. \qquad (4.124)$$

Following the calculation for the single-pump case, the nonlinear phase noise variance is given by

$$\begin{aligned}\sigma^2_{xpm,pc} &= <(\phi^{pc}_{xpm})^2> - <\phi^{pc}_{xpm}>^2 \\ &= \frac{9}{4}\gamma^2 L^2 \left(<(P_{p1}+P_{p2})^2> - (<P_{p1}>+<P_{p2}>)^2\right) \\ &= \frac{9}{4}\gamma^2 L^2 \left(<(\Delta P_{p1})^2> + <(\Delta P_{p2})^2> + 2<\Delta P_{p1}\Delta P_{p2}>\right) \\ &= 9\gamma^2 L^2 \left(<P_{p1}><\Re\{\Delta A_{p1}\}^2> + <P_{p2}><\Re\{\Delta A_{p2}\}^2>\right). \end{aligned} \quad (4.125)$$

where it was used that the noise contributions of the two pumps are independent of each other, meaning that $<\Delta A_{p1}\Delta A_{p2}>=0$. If the two pumps have the same average power, $<P_{p1}>=<P_{p2}>$ as well as equal pump SNR values, $SNR_{p1}=SNR_{p2}$, Eq. 4.125 can be rewritten to

$$\begin{aligned}\sigma^2_{xpm,pc} &= 9\gamma^2 L^2 \frac{<P_{p1}>^2}{SNR_{p1}} \\ &= 4\frac{<\phi^{pc}_{XPM}>^2}{SNR_{p1}}. \end{aligned} \quad (4.126)$$

Assuming perfect phase matching and a high gain/conversion efficiency and using $<P_{p1}>=<P_{p2}>$, the maximum gain/conversion efficiency for the phase-conjugation process given in Eq. 4.23 can be approximated by

$$\max\{G^{pc}_i\} \cong \max\{G^{pc}_s\} \cong \frac{1}{4}\exp(4\gamma <P_{p1}> L). \quad (4.127)$$

Then, the lower bound on the nonlinear phase noise variance for a FOPA with a certain gain/conversion efficiency is given by

$$\sigma^2_{xpm,pc} \cong \frac{9}{16}\frac{\ln^2(4\max\{G^{pc}_s\})}{SNR_{p1}} = \frac{9}{8}\sigma^2_{xpm,sp} \quad (4.128)$$

Comparison to Eq. 4.116 shows that the phase noise variance is slightly higher for the phase-conjugation process than for the single-pump process leaving all conclusions from the previous section stay also quantitatively valid. This is validated by Fig. 4.33 which shows $\sigma^2_{xpm,pc}$ as function of the maximum average gain and for different pump SNR. The solid lines are calculated with Eq. 4.128 while the symbols correspond to results from numerical simulations using the NLS equation (2.39) with a CW input signal and two CW pump signals. The used parameters were L = 1 km, $P_{p1} = P_{p2}$, P_s = -30 dBm, $\alpha = 0$, $\gamma = 10$(W km)$^{-1}$, $\lambda_{zd} = 1553$nm, $\lambda^{pc}_a - \lambda_{zd} = -0.05 nm$, $\lambda_{p1} - \lambda^{pc}_a = 25$ nm, $\beta_3 = 0.033$ ps^3/km and $\beta_4 = 2.5 \times 10^{-4}$ps^4/km, $\lambda_s - \lambda^{pc}_a = 15 nm$. The output signal was optically filtered by a 2nd order Gaussian bandpass filter with 25 GHz bandwidth before evaluating the noise variances. As for the single-pump configuration, the analytical results show a good agreement with the simulation. For completeness, Fig. 4.34 shows the (O)SNR penalty for different phase

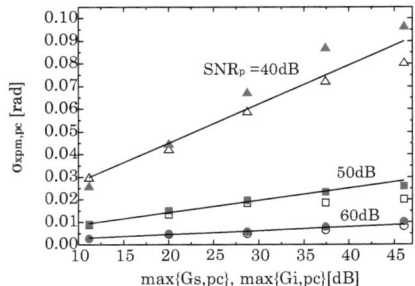

Figure 4.33: Standard deviation of the PC pump-induced XPM phase distortion for signal and idler as a function of the maximum gain and different pump SNR (solid line: theory after Eq. 4.116, open symbols: simulation results for the signal, filled symbols: simulation results for the idler. The used parameters are given in the text.

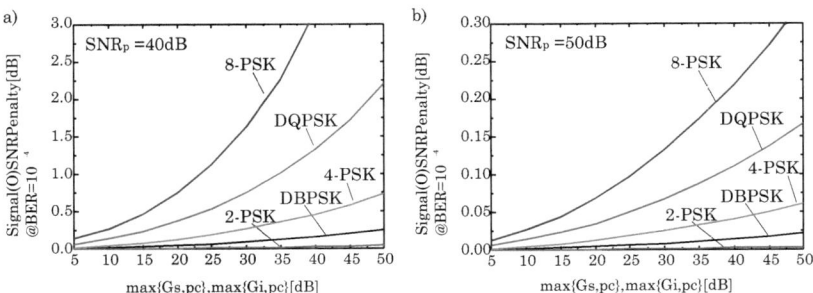

Figure 4.34: Signal (O)SNR penalty @ BER = 10^{-4} for different phase modulation formats as a function of $\max\{G_s^{pc}\}$, $\max\{G_i^{pc}\}$ ($P_{p1} = P_{p2}$) and a) $SNR_{p1} = SNR_{p2} = 40$ dB and b) $SNR_{p1} = SNR_{p2} = 50$ dB

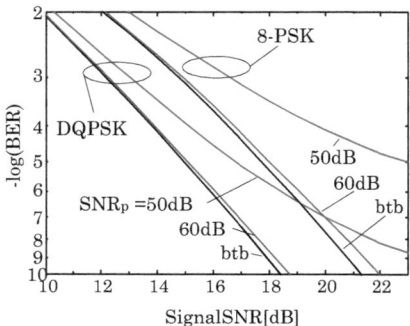

Figure 4.35: BER as a function of the signal SNR for DQPSK and 8-PSK after 10 cascaded FOPA stages with $\max\{G_s^{pc}\} = 40dB$ in the back-to-back case and for different values of the pump SNR ($P_{p1} = P_{p2}$, $SNR_{p1} = SNR_{p2}$).

modulation formats as a function of maximum gain /conversion efficiency. The BER was calculated by inserting the nonlinear phase noise variance given in Eq. 4.128 into Eqs. 3.32 and 3.34. Fig. 4.35 shows the BER for DQPSK and 8-PSK after 10 cascaded FOPA stages with $\max\{G_s^{pc}\} = 40dB$, i.e. with gain comparable to an EDFA. Similar requirements on the pump signal quality as for the single-pump configuration can be derived, i.e. a pump SNR of 50 dB is necessary for a negligible penalty after single-stage amplification while the necessary pump SNR increases to 60 dB for 10 cascaded FOPA stages.

4.4.3 Pump-Induced Amplitude Noise in the Single-Pump Configuration

Similarly to the pump-induced nonlinear phase noise, noisy pumps will also generate nonlinear amplitude noise because the gain and the conversion efficiency depends on the pump power [118, 120, 47]. As for the pump-induced phase noise, the generation of the nonlinear amplitude noise is pronounced only for high gain /conversion efficiency of the FOPA. The field gain /conversion efficiency of the single-pump FOPA is given by Eqs. H.21 and H.22 for perfect phase-matching and undepleted pump waves. In the high gain regime, both are approximately the same and can be written as

$$\mathscr{G}_s^{sp} \cong \frac{1}{2}\exp\left(\gamma|A_p|^2 L\right) \cong \mathscr{G}_i^{sp} \qquad (4.129)$$

Note that walk-off effects [130] are neglected in Eq. 4.129. If the pump wave is distorted by noise, Eq. 4.106 can be used and Eq. 4.129 can be rewritten as

$$\mathcal{G}_s^{sp} = \frac{1}{2}\exp(\gamma L <A_p>^2)\exp[2\gamma L <A_p>\Re\{\Delta A_p\}].$$

The mean field gain is defined by

$$<\mathcal{G}_s^{sp}> = \frac{1}{2}\exp(\gamma L <A_p>^2) \qquad (4.130)$$

and is related to the mean power gain by $<G_s^{sp}> = <\mathcal{G}_s^{sp}>^2$. The field gain fluctuation

$$\widetilde{\mathcal{G}}_s^{sp} = \exp[2\gamma L <A_p>\Re\{\Delta A_p\}]. \qquad (4.131)$$

is a multiplicative noise source representing gain fluctuations due to amplitude fluctuations of the pump waves. Since $\Re\{\Delta A_p\}$ is Gaussian distributed, $\widetilde{\mathcal{G}}_s^{sp}$ exhibits log-normal statistics. Its probability distribution function is given by

$$\text{PDF}_{\widetilde{\mathcal{G}}_s^{sp}}(x) = \begin{cases} \frac{1}{\sqrt{2\pi\sigma^2}x}\exp\left[-\frac{(\ln x)^2}{2\sigma^2}\right] & x > 0 \\ 0 & x \le 0 \end{cases} \qquad (4.132)$$

The parameter σ^2 is the variance of the argument of the exponential function in Eq. (4.131) and is given by

$$\sigma^2 = 4\gamma^2 L^2 <A_p>^2 <\Re\{\Delta A_p\}^2> \qquad (4.133)$$

$$= 2\gamma^2 L^2 \frac{P_p^2}{\text{SNR}_p} \qquad (4.134)$$

$$= \frac{\ln^2(4\max(G_s^{sp}))}{(2\,\text{SNR}_p)}.$$

where Eqs. 4.107 and 4.115 were used. Thus, the PDF of the gain fluctuations is completely determined by the FOPA maximal power gain and the pump SNR. The mean value of $\widetilde{\mathcal{G}}_s^{sp}$ is given by

$$\langle\widetilde{\mathcal{G}}_s^{sp}\rangle = \exp\left(\frac{1}{2}\sigma^2\right) \approx 1 \qquad (4.135)$$

since σ^2 is small. The variance of $\widetilde{\mathcal{G}}_s^{sp}$, i.e. the variance of the nonlinear amplitude noise, is given by

$$\sigma_{nan,sp}^2 = \exp(\sigma^2)(\exp(\sigma^2)-1)$$
$$\approx (1+\sigma^2)\sigma^2$$
$$\approx \sigma^2 = \sigma_{xpm,sp}^2. \qquad (4.136)$$

where $\sigma^2 \ll 1$ and $\exp(x) \approx 1+x$ for $x \ll 1$ were used. Thus, for the single-pump FOPA, the nonlinear amplitude noise variance has the same magnitude as the pump-induced nonlinear phase noise given in Eq. 4.116. Still it is to note that both noise contributions exhibit different probability distribution functions. In Fig. 4.36, the nonlinear amplitude noise standard deviation after

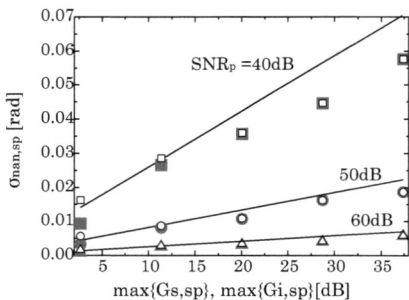

Figure 4.36: Standard deviation of the nonlinear amplitude noise for the amplified signal and the idler of a single-pump FOPA as a function of the maximum power gain and different pump SNR (solid line: theory after Eq. 4.136 (for both signal and, open symbols: simulation results for the signal, filled symbols: simulation results for the idler. The used parameters are given in the text.

Eq. 4.136 is shown in comparison to results of numerical simulations using Eq. 2.39 with a CW input signal and a CW pump signal. The used parameters were L = 1 km, P_s = -30 dBm, $\alpha = 0$, $\gamma = 10(\text{W km})^{-1}$, λ_{zd} = 1553nm, $\lambda_p - \lambda_{zd}$ = 1.1nm, β_3 = 0.033 ps^3/km and $\beta_4 = 2.5 \times 10^{-4}$ps^4/km. λ_s was adjusted to the gain peak for different pump powers. The output signal was optically filtered by a 2nd order Gaussian bandpass filter with 25 GHz bandwidth before evaluating the noise variances. The analytical results show a good agreement with the simulation.

As purely phase modulated signals are insensitive against amplitude noise, the impact of the nonlinear amplitude noise is shown by calculating (O)SNR penalties for the 16-QAM format that contains both amplitude and phase modulation. The BER calculation for QAM signals has to take into account both nonlinear amplitude and phase noise and is described in more detail in App. I. The resulting signal (O)SNR penalty for square 16-QAM is shown in Fig. 4.37 for two different pump SNR values. The comparison to PSK formats shows that square 16-QAM performs slightly worse than 8-PSK. Additionally, the (O)SNR penalties for the 16-QAM format resulting taking into account only pump-induced nonlinear phase noise, but not pump-induced nonlinear amplitude noise are shown in Fig. 4.37a with a dashed line. This curve indicates that nonlinear phase noise is the dominating distortion which was confirmed by recent experiments [131]. Fig. 4.38 depicts the BER for square 16-QAM after 10 conversions as a function of the signal SNR. Using Eqs. 4.54

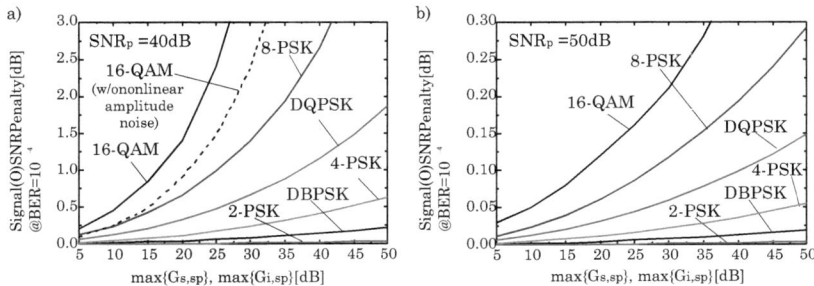

Figure 4.37: Signal (O)SNR penalty @ BER = 10^{-4} for 16-QAM and different phase modulation formats as a function of $\max(G_s^{sp})$ and a) SNR_p = 40 dB and b) SNR_p = 50 dB

and 4.56, the field gain fluctuation after N_c cascaded amplifications / wavelength conversions is given by

$$\widetilde{\mathcal{G}}_s^{sp}\Big|_{\Sigma N_c} = \prod_{l=1}^{N_c} \widetilde{\mathcal{G}}_{s,l}^{sp} = \prod_{l=1}^{N_c} \exp\left[2\gamma L <A_p> \Re\{\Delta A_{p,l}\}\right]. \quad (4.137)$$

Using Eqs. 4.136 and 4.133,

$$\sigma_{nan,sp}^2\Big|_{\Sigma N_c} = \sum_{l=1}^{N_c} \sigma_{nan,sp}^2\Big|_l. \quad (4.138)$$

This is the same result as for the pump-induced nonlinear phase noise given in Eq. 4.117, i.e., for equal single stages, the accumulated nonlinear amplitude noise variance equals $N_c \times \sigma_{nan,sp}^2$. Fig. 4.38 shows that penalty free amplification and wavelength conversion of 16-QAM puts similar requirements on the pump noise as 8-PSK.

4.4.4 Pump-Induced Amplitude Noise in the Dual-Pump Configuration

For the dual-pump setup, the BER calculation in the presence of pump-induced amplitude noise is similar to the single-pump setup. Furthermore, also here, only the phase-conjugation process is taken into account because the effect is negligible for the frequency-conversion process. The field gain/conversion efficiency of the dual-pump phase-conjugation based FOPA is given in the case of perfect phase-matching and undepleted pump waves by Eqs. H.46 and H.47 and can be approximated in the high gain regime as

$$\mathcal{G}_s^{pc} \cong \frac{1}{2}\exp\left(2\gamma|A_{p1}||A_{p2}|L\right) \cong \mathcal{G}_i^{pc}. \quad (4.139)$$

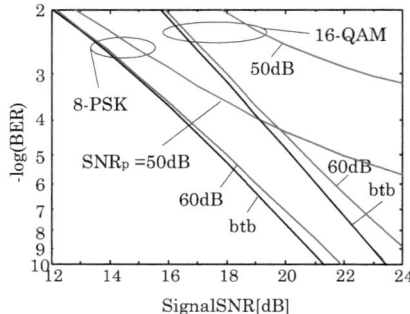

Figure 4.38: BER as a function of the signal SNR for 16-QAM and 8-PSK after 10 cascaded AOWCs stages with $\max(G_s^{sp}) = 40$ dB in the back-to-back case and for different values of the pump SNR

Using Eqs. 4.121,

$$\begin{aligned}|A_{p1}||A_{p2}| &= \sqrt{|A_{p1}|^2|A_{p2}|^2}\\ &\cong <A_{p1}><A_{p2}>\sqrt{1+2\frac{<A_{p2}>}{\Re\{\Delta A_{p1}\}}+2\frac{<A_{p1}>}{\Re}\{\Delta A_{p2}\}}\\ &\cong <A_{p1}><A_{p2}>+<A_{p1}>\Re\{\Delta A_{p2}\}+<A_{p2}>\Re\{\Delta A_{p1}\}.\end{aligned} \quad (4.140)$$

By inserting Eq. 4.140 into Eq. 4.139, the mean field gain can defined as

$$<\mathcal{G}_s^{pc}>\cong \frac{1}{2}\exp\left(2\gamma L<A_{p1}><A_{p2}>\right) \quad (4.141)$$

and is related to the mean power gain $<G_s^{pc}>=<\mathcal{G}_s^{pc}>^2$. The field gain fluctuation

$$\widetilde{\mathcal{G}}_s^{pc}=\exp\left[2\gamma L(<A_{p1}>\Re\{\Delta A_{p2}\}+<A_{p2}>\Re\{\Delta A_{p1}\})\right]. \quad (4.142)$$

also has a log-normal distribution as given by Eq. 4.132, but with a parameter

$$\begin{aligned}\sigma^2 &= 4\gamma^2L^2\left(<A_{p1}>^2<\Re\{\Delta A_{p2}\}^2>+<A_{p2}>^2<\Re\{\Delta A_{p1}\}^2>\right)\\ &= 2\gamma^2L^2\left(\frac{P_{p1}^2}{\text{SNR}_{p1}}+\frac{P_{p2}^2}{\text{SNR}_{p2}}\right)\\ &= 4\gamma^2L^2\frac{P_{p1}^2}{\text{SNR}_{p1}}\\ &= \frac{\ln^2(4\max\{G_s^{PC}\})}{(4\,\text{SNR}_p)},\end{aligned} \quad (4.143)$$

where Eqs. 4.107 and 4.127 were used. Additionally, equal pump powers, $<P_{p1}>=<P_{p2}>$, and equal pump SNRs, $\text{SNR}_{p1}=\text{SNR}_{p2}$, were assumed. As for the single-pump FOPA, the parameter σ^2 is approximately equal to the

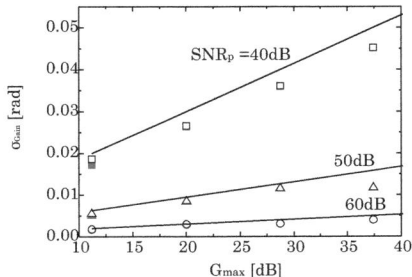

Figure 4.39: Standard deviation of the pump-induced nonlinear amplitude noise for signal and idler of the phase-conjugation process as a function of the maximum gain and different pump SNR (solid line: theory after Eq. 4.143 , open symbols: simulation results for the signal, filled symbols: simulation results for the idler. The used parameters are given in the text.

variance $\sigma_{nan,pc}^2$ of the gain fluctuation $\widetilde{\mathcal{G}}_s^{pc}$, i.e. the variance of the nonlinear amplitude noise. It is again completely determined by the FOPA mean power gain and the pump SNR. In Fig. 4.39, the standard deviation of the nonlinear amplitude noise is shown in comparison to results of the numerical simulations using Eq. 2.39 with a CW input signal and two CW pump signals. The used parameters were L = 1 km, $P_{p1} = P_{p2}$, P_s = -30 dBm, $\alpha = 0$, $\gamma = 10$(W km)$^{-1}$, λ_{zd} = 1553nm, $\lambda_a^{pc} - \lambda_{zd}$ = -0.05nm, $\lambda_{p1} - \lambda_a^{pc}$ = 25 nm, $\beta_3 = 0.033$ ps^3/km and $\beta_4 = 2.5 \times 10^{-4}$ps^4/km, $\lambda_s - \lambda_a^{pc}$ = 15 nm. The output signal was optically filtered by a 2nd order Gaussian bandpass filter with 25 GHz bandwidth before evaluating the noise variances. The BER calculations after appendix I given in Figs. 4.40 and 4.41 are very similar to the single-pump case such that the conclusions are the same. In particular, pump-induced nonlinear phase noise can be again identified as the dominating distortion in comparison to the pump-induced nonlinear amplitude noise.

4.5 Signal-Induced Phase Noise

In the following section, nonlinear phase noise induced by the amplitude noise of the signal itself is treated which affects both the non-converted output signal and the idler. It is taken into account by ϕ_{spm} in the Eqs. 4.44, 4.45, 4.49, 4.50, 4.52 and 4.53. The physical origin of the phase noise generation is SPM and XPM as discussed in section 2.1.3 and 2.1.3. This effect cannot be treated analytically with the approximate equations given in Appendix H because the

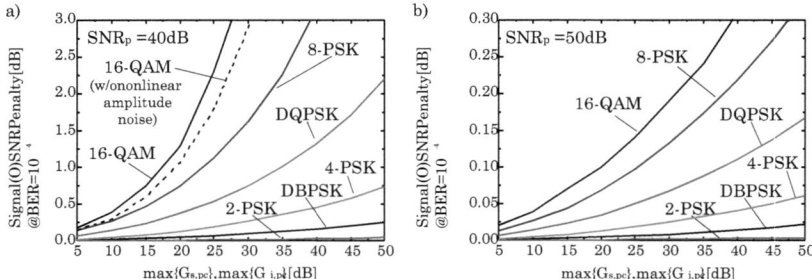

Figure 4.40: Signal (O)SNR penalty @ BER = 10^{-4} for 16-QAM and different phase modulation formats as a function of the maximum parametric power gain $\max\{G_s^{PC}\}$ ($P_{p1} = P_{p2}$) and a) $\text{SNR}_{p1} = \text{SNR}_{p2} = 40$ dB and b) $\text{SNR}_{p1} = \text{SNR}_{p2} = 50$ dB

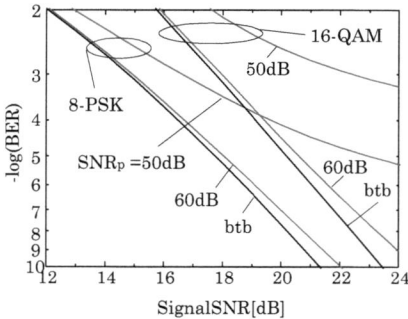

Figure 4.41: BER as a function of the signal SNR for 16-QAM and 8-PSK after 10 cascaded FOPA stages with $\max\{G_s^{PC}\} = 40dB$ in the back-to-back case and different pump SNR ($P_{p1} = P_{p2}$, $\text{SNR}_{p1} = \text{SNR}_{p2}$)

underlying differential equations cannot be linearized in this case [14]. So, only results of numerical simulations will be presented. In these simulations, the pump-phase modulation as well as pump noise is neglected and only the signal noise is taken into account. Similarly to the previous section, direct and coherent detection formats will be treated in parallel.

4.5.1 Single-Pump Configuration

The simulation setup and the results for the single-pump configuration are shown in Fig. 4.42. To study the generation of signal-induced nonlinear phase noise independently from other effects discussed in the sections above, the CW input signal is distorted with amplitude noise and is inserted in the HNLF together with a noise-free and unmodulated pump signal [15]. At the output, the output signal and idler magnitude and noise is evaluated. The used parameters were L = 1 km, $\alpha = 0$, $\gamma = 10 (\text{W km})^{-1}$, $\lambda_{zd} = 1553$nm, $\lambda_p - \lambda_{zd} = 1.1$ nm, $\beta_3 = 0.033$ ps^3/km and $\beta_4 = 2.5 \times 10^{-4}$ps^4/km. λ_s was adjusted to the gain peak for different pump powers. The output signal was optically filtered by a 2nd order Gaussian bandpass filter with 25 GHz bandwidth before evaluating the noise variances. Fig. 4.42b shows the signal gain and the conversion efficiency as a function of the pump power. As a second parameter, the signal input power was varied from -30 dBm to 0dBm. For low input powers, the gain always takes its maximum value, i.e., there is no gain saturation. For high signal input powers the gain decreases due to pump depletion. Fig. 4.42c and d show the output signal and idler amplitude standard deviation normalized to the mean for the same parameter set as in Fig. 4.42b. For low signal input powers, the output amplitude fluctuations equal the input fluctuations, i.e., the FOPA acts as a linear amplifier. For high signal input powers, the output amplitude fluctuations decrease. In this regime, the gain is saturated and the FOPA acts like a limiting amplifier suppressing amplitude fluctuations. However, the FOPA is not anymore transparent for amplitude modulation formats like OOK and QAM in this regime. Fig. 4.42e and f show the output signal and idler phase standard deviations for the same parameter set as in Fig. 4.42b. For low signal input powers, the generated phase noise is very small. Only for input signal powers leading to gain saturation, i.e. to nonlinear amplification, the phase noise is higher although still small even for an input signal SNR of 20 dB. Comparison to the previous section shows that

---

[14]Very recently, an analytical approach for the approximate treatment of saturated FOPAs was presented [132].

[15]Stimulated Brillouin scattering is neglected.

the standard deviations of the generated noise will lead to negligible (O)SNR penalties. Thus, by choosing sufficiently small signal input powers both full transparency of the FOPA for amplitude modulation formats like OOK and QAM and, at the same time, negligible phase distortions due to signal SPM and XPM can be achieved. As a rule of thumb, the signal output power should be 10 dB below the pump power.

4.5.2 Dual-Pump Configuration

For the dual-pump configuration based on the PC process, the simulation setup and the results are shown in Fig. 4.43. Here, the amplitude-noise distorted CW input signal is combined with two clean and unmodulated pump signals. The used parameters were L = 1 km, $P_{p1} = P_{p2}$, $\alpha = 0$, $\gamma = 10 (\text{W km})^{-1}$, $\lambda_{zd} = 1553$nm, $\lambda_p - \lambda_{zd} = -0.05 nm$, $\lambda_{p1} - \lambda_a^{pc} = 25$ nm, $\beta_3 = 0.033$ ps^3/km and $\beta_4 = 2.5 \times 10^{-4}$ps^4/km, $\lambda_s - \lambda_a^{pc} = 15 nm$. The output signal was optically filtered by a 2nd order Gaussian bandpass filter with 25 GHz bandwidth before evaluating the noise variances. A comparison to Fig. 4.42 yields that the dual-pump configuration shows a similar behavior as the single-pump configuration. Thus, also in this case, a signal output power which is 10 dB below the pump powers will ensure full transparency and negligible signal-induced phase distortions at the same time. A similar conclusion is valid for the FOPA based on the frequency-conversion process.

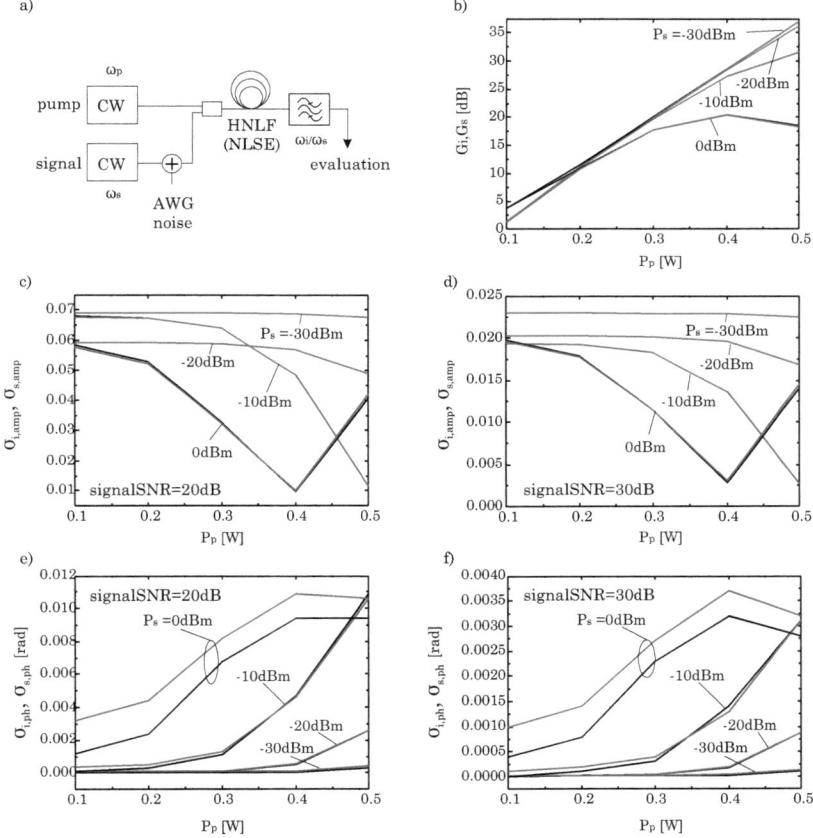

Figure 4.42: a) Simulation setup for the characterization of the SPM phase distortions in the SP-based FOPA, b) signal gain and conversion efficiency, c) output signal and idler amplitude standard deviation (normalized to mean) for signal SNR of 20dB, d) same as c) but for signal SNR of 30dB, e) output signal and idler phase standard deviation for signal SNR of 20 dB, f) same as e) but for signal SNR of 30 dB (red - idler, black - signal)

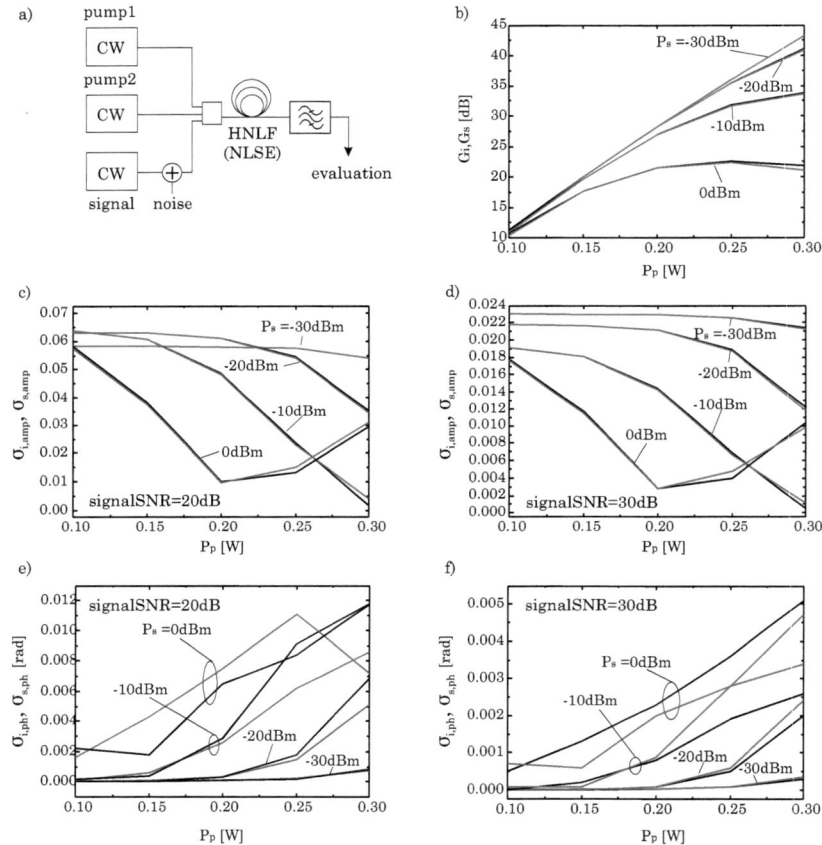

Figure 4.43: a) Simulation setup for the characterization of the SPM phase distortions in the PC-based FOPA, b) signal gain and conversion efficiency, c) output signal and idler amplitude standard deviation (normalized to mean) for signal SNR of 20dB, d) same as c) but for signal SNR of 30dB, e) output signal and idler phase standard deviation for signal SNR of 20 dB, f) same as e) but for signal SNR of 30 dB (red - idler, black - signal)

Chapter 5

Wavelength Converters Based on Four-Wave Mixing in SOA

In this chapter, wavelength converters based on FWM in SOA are discussed in a very similar way as the FOPAs in chapter 4. In difference to the previous chapter, all quantitative results are obtained by the numerical SOA model presented in chapter 2 because the analytical description of the SOA is much more difficult than that for the HNLF. As before, the individual phase distortions are discussed in detail and their impact on the BER of various phase-modulation formats is quantitatively given.

5.1 General Characteristics

5.1.1 Setup

In principle, the same FWM processes occur in the SOA as those discussed for the HNLF in section 4.1.1, i.e., one generally can distinguish between degenerate and non-degenerate FWM, both described in section 2.1.3. The setup of the single-pump configuration relying on degenerate FWM is shown in Fig. 5.1 [133, 134, 135, 136]. The (weak) input signal is combined with a single (strong) pump wave and fed into the SOA. A single converted signal (called idler in the following) is generated by the degenerate FWM which is filtered out by a bandpass filter. The nonlinear process is characterized by an energy transfer from the pump to the signal and the idler. Thus, parametric amplification of the signal is possible in this scheme (although typically not reached in SOA-based AOWC for reasons explained later in more detail). With Eq. 2.36, the idler frequency is given by

$$\omega_i = 2\omega_p - \omega_s. \tag{5.1}$$

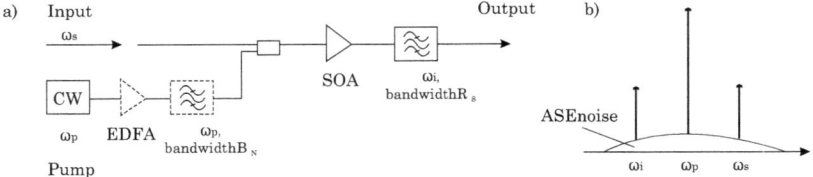

Figure 5.1: a) Single-pump configuration of the SOA-based AOWC, b) schematic SOA output spectrum

Since the dispersion in the SOA is generally negligible until very long SOAs and very large signal-pump wavelength detunings are used [66] the phase matching condition given in Eq. 2.37,

$$\Delta B_{sp} = 2B_p - B_s - B_i \approx 0, \qquad (5.2)$$

is almost always fulfilled. This makes the SOA-based wavelength converters fully tunable within the gain bandwidth of the SOA [1], in contrast to the HNLF-based converters. As a second difference to the HNLF-based wavelength converters, no phase modulation of the pump is needed since Brillouin scattering can be neglected due to the short length of the SOA making the setup of the SOA-based wavelength converter less complex. However, since the SOA is an active device, it generates an amplified spontaneous emission noise floor that leads to an increased noise figure in comparison to the HNLF-based converters as will be shown in the next sections.

Dual-pump configurations can be also used in the SOA in order to provide a wavelength-independent [137] or polarization-independent conversion efficiency [138, 139]. However, in this thesis, only the single-pump configuration will be treated because the focus lies on the analysis of phase distortions. The dual-pump configuration for HNLF-based wavelength converters enables the suppression of the pump-phase modulation (see section 4.3.2) which presents a key advantage against the single-pump configuration and justifies the extensive investigations. Since a pump-phase modulation is generally not needed for SOA-based wavelength converters, no qualitative differences are expected from the dual-pump configuration in this case (as may also be estimated from the rather similar results on pump- and signal-induced noise for the single- and dual-pump configuration based on HNLF treated in the sections 4.4 and 4.5).

In the following, all simulations will be performed with the SOA model

---

[1]Outside the gain bandwidth, absorption of the interacting wave hinders efficient FWM.

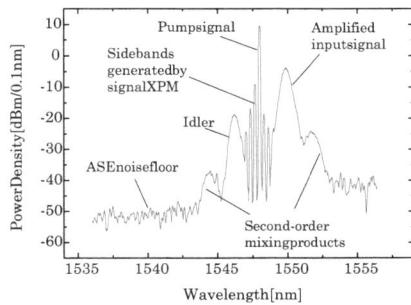

Figure 5.2: Simulated output spectrum (before the output bandpass filter) of the SOA-based single-pump wavelength converter (40 Gb/s DQPSK input signal, L = 1 mm, I_B = 190 mA, P_p = 12.8 dBm, P_s = 2.8 dBm)

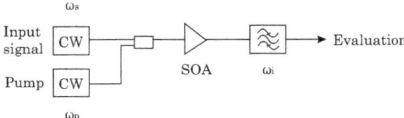

Figure 5.3: Simulation setup to characterize the conversion efficiency, the output OSNR and the noise figure of the SOA-based SP AOWC

presented in section 2.3. The changing simulation parameters are given in the text while a summary over all SOA parameters is given in section G together with corresponding simulated gain and ASE curves. Fig. 5.2 shows a simulated output spectrum (before the output bandpass filter) of the SOA-based single-pump wavelength converter. The comparison with corresponding experimental results [140] validates that the SOA model reproduces all features of the experiment with high accuracy such as the ASE noise floor, the second-order mixing products and the additional signal-induced XPM of the pump wave.

5.1.2 Conversion Efficiency

In Fig. 5.4a, the conversion efficiency G_i of the single-pump configuration as defined by Eq. 4.14 is shown as a function of the signal-pump detuning given by

$$\Delta \lambda = \lambda_p - \lambda_s = 2\pi c_0 \left(\frac{1}{\omega_p} - \frac{1}{\omega_s} \right) \cong \frac{2\pi c_0}{\omega_p^2} (\omega_s - \omega_p) \tag{5.3}$$

with $\Delta \lambda \ll \lambda_p$. As follows from Fig. 5.3, $|2\Delta\lambda|$ equals the conversion range $|\lambda_i - \lambda_s|$. The used parameters were L = 1 mm, I_B = 190 mA, P_p = 12.8 dBm and P_s = 2.8 dBm. λ_p was set at the ASE spectral peak. For signal-pump

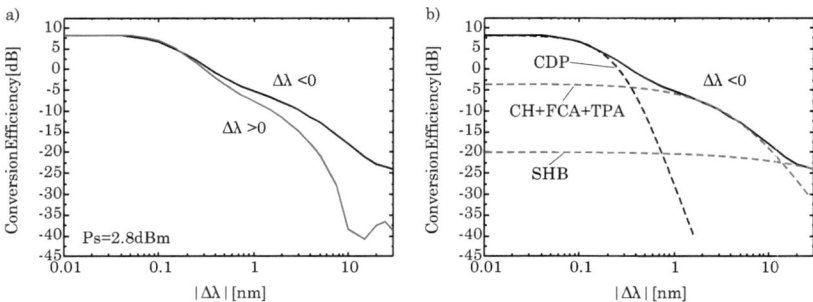

Figure 5.4: a) Simulated conversion spectrum of the single-pump SOA-based AOWC, b) conversion spectrum with schematic depiction of the individual contributions of the underlying nonlinear effects. The used parameters are given in the text.

detunings smaller than 0.3 nm, conversion efficiencies larger than 1 can be achieved. However, for larger $|\Delta\lambda|$, the conversion efficiency decreases very quickly. In comparison to the conversion spectrum for the single-pump process in the HNLF shown in Fig. 4.4, the conversion spectrum for the SOA-based converter is extremely narrow. The reason for this are the rather slow resonant nonlinearities in the SOA (in comparison the fs nonlinear response of the HNLF). As shown schematically in Fig. 5.4b, all nonlinear effects described in section 2.3.3 contribute to the conversion spectrum. The shape of the constituents are low pass filter-like with different strengths and bandwidths determined by the time constants of the nonlinear effects [141]. Therefore, CDP with a time constant of several 10 ps is dominating up to a signal-pump detuning of 0.5 nm, while CH, FCA and TPA with time constants of about 1 ps dominate from 1 nm to 10 nm. For $|\Delta\lambda| > 10$ nm, the main contribution comes from SHB with a time constant of about 100 fs. Furthermore, the conversion spectrum is not symmetric. This is due to the simultaneous presence of gain and index gratings related to each nonlinear effect [64]. The phase offset between them is dependent on the alpha factor and leads to constructive or destructive interference depending whether the signal is situated on the short- or long-wavelength side of the pump. The dependence of the conversion efficiency on the pump power is shown in Fig. 5.5 a) and b) for different fixed input signal powers and signal-to-pump power ratios, respectively. The used simulation parameters were L = 1 mm, I_B = 190 mA and $\Delta\lambda = -2.5$ nm. λ_p was set to the ASE spectral peak. To understand the graphs it is important to note that G_i depends on the FWM conversion efficiency as well as on the SOA gain. Because the former grows with the pump power

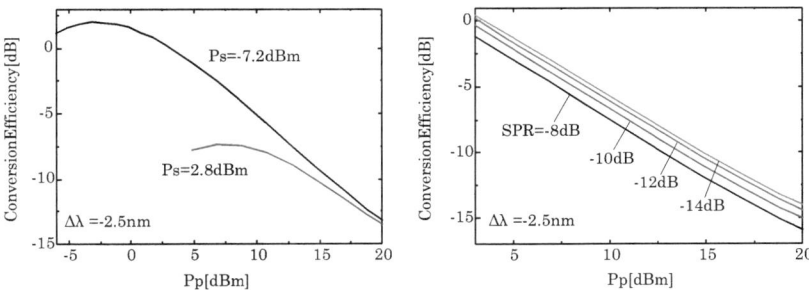

Figure 5.5: a) Conversion efficiency of the single-pump SOA-based AOWC as a function of the pump power and for different fixed input signal powers, b) same as a) but for different fixed signal-to-pump power ratios. The used parameters are given in the text.

while the latter decreases with the pump power due to the gain saturation, there exists a maximum G_i for a fixed input power. This is another major difference to the HNLF-based wavelength converters for which the conversion efficiency (for a small and fixed signal power) is always growing with the pump power as confirmed by Figs. 4.4, 4.6 and 4.7. For a fixed signal-to-pump power ratio (SPR), the relative contributions of FWM and SOA gain remain constant and G_i is monotonically decreasing due to the increasing gain saturation of the SOA. The dependence of the conversion efficiency on the length of the SOA is shown in Fig. 5.6 for both a fixed signal power and a fixed SPR. The used simulation parameters were I_B = 190 mA/mm and $\Delta\lambda = -2.5$ nm. λ_p was set to the ASE spectral peak which changes for the different lengths as given by Tab. G.2. G_i grows with the length since the interaction length of the participating waves increases. Fig. 5.7a shows that the conversion efficiency is only weakly dependent on the pump wavelength. Here, the used simulation parameters were L = 1 mm, I_B = 190 mA, P_p = 12.8 dBm, P_s = 2.8 dBm and $\Delta\lambda = -2.5$ nm. In Fig. 5.7b, it is shown that the conversion efficiency grows with the pump current. However, for high I_B, the SOA suffers from thermal problems effectively limiting the enhancement of the conversion efficiency due to this approach in practice. The used simulation parameters were L = 1mm, P_s = 2.8 dBm and $\Delta\lambda = -2.5$ nm.

5.1.3 Noise Figure

The noise figure of the SOA-based AOWC can be defined in the same way given in Eq. 4.29 for the HNLF-based AOWC. The input SNR is limited by

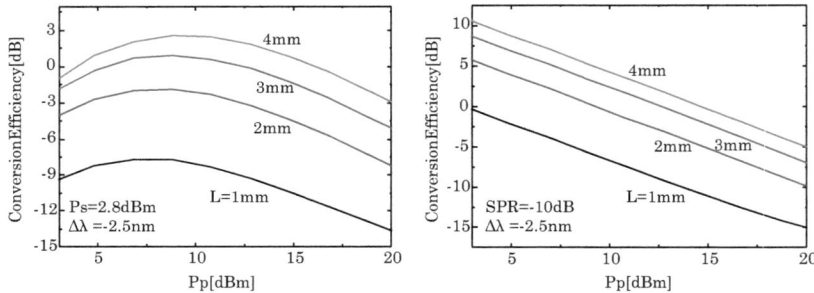

Figure 5.6: a) Conversion efficiency of the single-pump SOA-based AOWC as a function of the pump power and for different SOA lengths and for a constant signal input power of 2.8 dB, b) same as a) but for a constant SPR of -10 dB. The used parameters are given in the text.

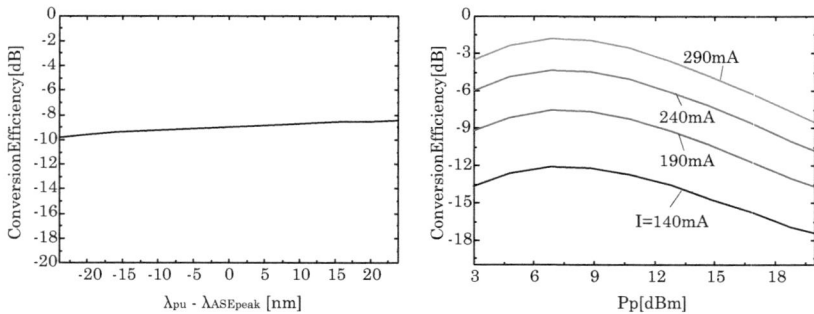

Figure 5.7: a) Conversion efficiency of the single-pump SOA-based AOWC as a function of the pump wavelength relative to the ASE peak and b) Conversion efficiency of the single-pump SOA-based AOWC as a function of the pump power and for different pump currents. The used parameters are given in the text.

128

quantum noise given by Eq. 4.30. However, the output SNR is typically dominated by the ASE noise spectral density in one polarization at the idler wavelength, $\rho_{ASE}(\omega_i)$. Thus, it is given by [142]

$$\mathrm{SNR}_{out} = \frac{G_i P_s}{\rho_{ASE}(\omega_i) R_s}. \tag{5.4}$$

Then, the noise figure is given by

$$\mathrm{NF}_i = \frac{2\rho_{ASE}(\omega_i)}{(\hbar \omega_i) G_i}. \tag{5.5}$$

Note that Eq. 5.5 does not take into account input and output coupling losses which lead to a further increase of the noise figure. Fig. 5.8a shows the noise figure as a function of $\Delta\lambda$. The used parameters were L = 1 mm, I_B = 190 mA, P_p = 12.8 dBm and P_s = 2.8 dBm. λ_p was set to the ASE spectral peak. In particular for signal-pump detunings above 1 nm, high noise figures above 20 dB occur. This is in parts due to the low conversion efficiencies shown in Fig. 5.4. Also for the HNLF-based wavelength converters, the noise figure increases if the conversion efficiency decreases (see Fig. 4.8). However, for the same conversion efficiency, the noise figure in the SOA-based converter is still much higher because of the additional ASE noise floor. Fig. 5.8b shows the noise figure as a function of the pump power and for different signal powers. The noise figure is increasing with the pump power because the conversion efficiency is decreasing. Furthermore, the noise figure is dependent on the signal unless the signal-to-pump power ratio gets to low values where the signal does not contribute to the SOA saturation. Because the noise figure should not depend on the signal power, its definition is, strictly speaking, meaningful only in this regime. The used simulation parameters were L = 1mm, I_B = 190 mA and $\Delta\lambda$ = −2.5 nm. λ_p was set to the ASE spectral peak. Fig. 5.9a shows the noise figure of the single-pump SOA-based AOWC as a function of the pump power and for different values of the SPR. The used simulation parameters were L = 1 mm, I_B = 190 mA and $\Delta\lambda$ = −2.5 nm. λ_p was set to the ASE spectral peak. The noise figure increases with the pump power and shows only a weak dependence on the signal-to-pump power ratio as expected because all chosen values for the SPR are small and the signal contributes only weakly to the SOA saturation. Fig. 5.9b shows the corresponding idler output OSNR (B_{ref} = 12.5 GHz). It is increasing with the pump power since the SOA is stronger saturated and the gain as well as the ASE power is decreasing. However, looking back at Fig. 5.5b confirms that the noise figure nevertheless increases because the conversion efficiency also decreases with the gain. As for the conversion efficiency, increasing the SOA length is beneficial also for the noise performance of the SP SOA AOWC as shown in Fig. 5.10. The

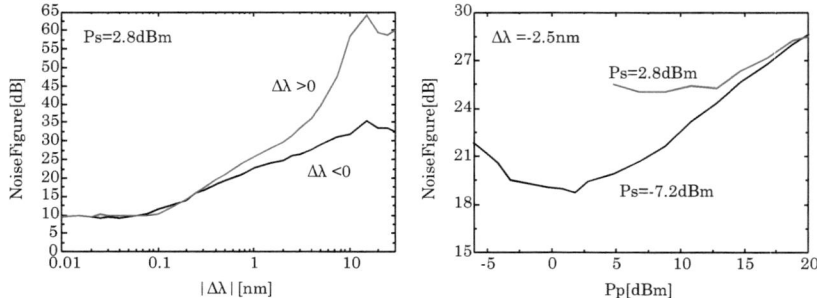

Figure 5.8: a) Noise figure of the single-pump SOA-based AOWC as a function of the conversion bandwidth and b) noise figure of the single-pump SOA-based AOWC as a function of the pump power and for different signal input powers. The used parameters are given in the text.

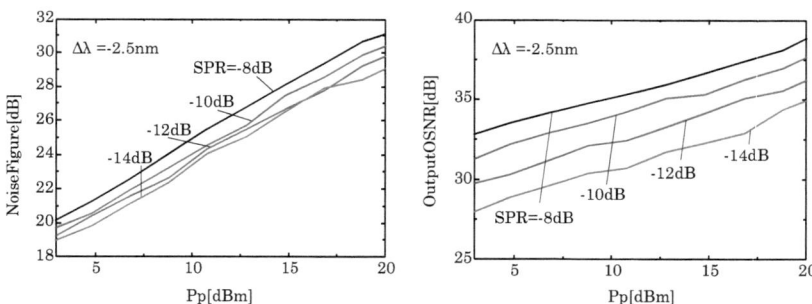

Figure 5.9: a) Noise figure and b) idler output OSNR (B_{ref} = 12.5 GHz) of the single-pump SOA-based AOWC as a function of the pump power and for different values for the SPR. The used parameters are given in the text.

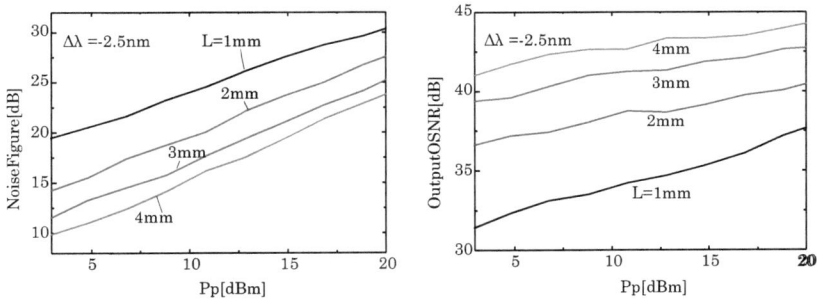

Figure 5.10: a) Noise figure and b) idler output OSNR (B_{ref} = 12.5 GHz) of the single-pump SOA-based AOWC as a function of the pump power and for different SOA lengths. The used parameters are given in the text.

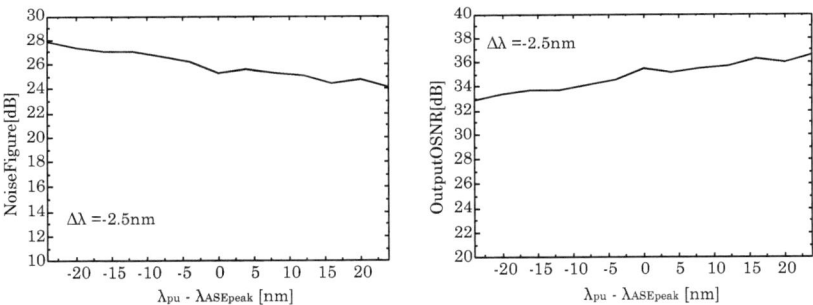

Figure 5.11: a) Noise figure and b) idler output OSNR (B_{ref} = 12.5 GHz) of the single-pump SOA-based AOWC as a function of the wavelength position of the pump wave relative to the ASE peak wavelength. The used parameters are given in the text.

used simulation parameters were I_B = 190 mA/mm and $\Delta\lambda$ = -2.5 nm. λ_p was set to the ASE spectral peak which changes for the different lengths as given by Tab. G.2. An increase of the SOA pump current has a similar effect. Finally, Fig. 5.11 shows the NF and the output OSNR (B_{ref} = 12.5 GHz) for different wavelength positions of the pump wave relative to the ASE peak wavelength. As shown before, the highest output OSNR is obtained on the longer wavelengths side of the ASE peak while the noise figure decreases to longer wavelengths. Here, the used simulation parameters were L = 1 mm, I_B = 190 mA, P_p = 12.8 dBm, P_s = 2.8 dBm and $\Delta\lambda$ = -2.5 nm.

5.1.4 Phase Distortions

Due to the equivalence of the nonlinear processes in the SOA and the HNLF, one can assume that the same additional phase distortions occur. They have been derived for the HNLF-based wavelength converters in Eqs. 4.44, 4.49 and 4.52. In full agreement, the different contributions to the idler phase distortion in the SOA-based converter can be written as

$$\Delta\phi_i = \phi_{lpn} + \phi_{xpm} + \phi_{spm}. \tag{5.6}$$

Here, ϕ_{lpn} is the contribution of the pump laser phase noise, ϕ_{xpm} is the contribution of the pump-amplitude noise which is transferred to idler phase noise by XPM, and ϕ_{spm} is the contribution of the signal amplitude noise which is transferred to idler phase noise by SPM and XPM. SPM and XPM are described in more detail in section 2.1.3 and 2.1.3. In contrast to the HNLF, no pump-phase modulation is used so that this contribution does not occur for the SOA-based AOWC.

5.2 Laser Phase Noise

Since the single-pump SOA-based wavelength converter relies on degenerate FWM, all conclusions drawn in section 4.2.1 for the single-pump HNLF-based converter are valid also in this case. In particular, the requirements on the pump laser linewidth are given in Fig. 4.11.

5.3 Impact of Pump-Induced Noise

Similarly to pump-induced noise in the HNLF-based AOWC, a noisy pump in the SOA-based AOWC will also generate nonlinear phase noise by XPM due to the presence of the alpha factor defined in Eqs. 2.67 and 2.72 [48]. This is accounted for by ϕ_{xpm} in Eq. 5.6. Although the mechanism for the generation of pump-induced nonlinear phase noise is the same in the SOA and the HNLF, a full analytical description as in the HNLF is not possible in the SOA due to its complicated saturation behavior. However, one can make an analytical estimate that qualitatively explains the results as will be seen in the next section.

5.3.1 Pump-Induced Phase Noise: Analytical Estimation

As shown in Fig. 5.1, the input signal wave shall be injected into the SOA together with a pump wave,

$$A(z,T) = A_p e^{i(B_p z - \Omega_p T)} + A_s e^{i(B_s z - \Omega_s T)}. \tag{5.7}$$

Thereby, the pump power shall much higher than the power of the input signal power,

$$|A(z,T)|^2 \cong |A_p|^2 = P_p. \tag{5.8}$$

To derive the nonlinear phase shift that depends on the gain via the alpha factor, first the nonlinear gain has to be calculated. Starting point is Eq. 2.88 describing the carrier dynamics in the SOA. The analytical estimate needs several simplifying assumptions that will be discussed in the following. First, the pump and the input signal are assumed to be of CW or quasi-CW type. This allows to set the time derivatives to zero. Second, the gain shall be independent of the wavelength. With Eq. 2.77, $g \cong g_p$. Third, the fast intraband effects are neglected, only CDP is considered which leads to $g_p = g_{\mathrm{CDP}}$ with Eq. 2.75. Fourth, only one polarisation and only the wave traveling in +z-direction is taken into account. With Eq. 2.89 follows that $(g \cdot S) = g_{\mathrm{CDP}} S^+$. Fifth, the SOA is assumed to be a lumped element, i.e. only one segment ($\Delta z \equiv L$) is taken into account and the carrier density as well as the optical power are constant over the SOA length. Together with all simplifications, Eq. 2.88 can be written as

$$0 = \frac{I_B}{q w_w d_w L} - R(N) - v_G g_{\mathrm{CDP}} S^+. \tag{5.9}$$

Additionally, the recombination term shall take the form $R(n) = N/\tau_s$ with τ_s the carrier lifetime. Using Eqs. 2.91 and 5.8, $S^+ \cong P_p/k_p$. Then, 5.9 can be further simplified to

$$\underbrace{\frac{I_B \tau_s}{q w_w d_w L}}_{N_{un}} = N + v_G g_{\mathrm{CDP}} \tau_s \frac{P_p}{k_p}. \tag{5.10}$$

Here, the unsaturated carrier density N_{un} was defined. Using Eq. 2.76, one can write

$$\underbrace{a_N[(N_{un} - N_{tr}) - (N - N_{tr})]}_{g_{\mathrm{CDP}}(N_{un}) - g_{\mathrm{CDP}}} = \underbrace{a_N v_G \tau_s}_{k_p/P_{sat}} g_{\mathrm{CDP}} \frac{P_p}{k_p}. \tag{5.11}$$

$g_{\mathrm{CDP}}(N_{un})$ is the unsaturated SOA gain and P_{sat} is the saturation input power. Rearranging gives finally

$$g_{\mathrm{CDP}}(P_p) = \frac{g_{\mathrm{CDP}}(N_{un})}{1 + P_p/P_{sat}}, \tag{5.12}$$

i.e the SOA gain effectively depends on the pump power. Now, the related phase change can be calculated by inspecting Eq. 2.84. Applying all approximations in this section and additionally neglecting spontaneous emission noise gives the propagation equation for the input signal wave,

$$A_s(L) = A_s(0)\exp\left\{\left(\frac{\Gamma}{2}g_{\text{CDP}}(P_p)(1+i\alpha_{H,\text{CDP}}) - \frac{a_{int}}{2}\right)L\right\} \quad (5.13)$$

Thus, the phase shift of the output signal wave is given by

$$\vartheta_s = \Im\left\{\ln\left(\frac{A_s(z)}{A_s(0)}\right)\right\} = \frac{\Gamma}{2}\alpha_{H,\text{CDP}}g_{\text{CDP}}(P_p)L$$
$$= \underbrace{\frac{\Gamma}{2}\alpha_{H,\text{CDP}}g_{\text{CDP}}(N_{un})L}_{\vartheta_{un}}\frac{1}{1+P_p/P_{sat}}. \quad (5.14)$$

Here, the phase shift for the unsaturated gain, ϑ_{un}, was defined. The pump wave shall exhibit an amplitude distortion, $P_p = <P_p> + \Delta P_p$, as given by Eq. 4.106. Inserting in Eq. 5.14 yields

$$\vartheta_s = \vartheta_{un}\left(1 + \frac{<P_p>}{P_{sat}} + \frac{<P_p>}{P_{sat}}\frac{\Delta P_p}{<P_p>}\right)^{-1}$$
$$= \vartheta_{un}\left(1 + \frac{<P_p>}{P_{sat}}\right)^{-1}\left(1 + \frac{\Delta P_p}{P_{sat} + <P_p>}\right)^{-1}$$
$$\cong \vartheta_{un}\left(1 + \frac{<P_p>}{P_{sat}}\right)^{-1}\left(1 - \frac{\Delta P_p}{P_{sat} + <P_p>}\right)$$
$$= \vartheta_{un}\left(\frac{1}{1+<P_p>/P_{sat}} - \frac{\Delta P_p/<P_p>}{P_{sat}/<P_p> + 2 + <P_p>/P_{sat}}\right) \quad (5.15)$$

where $(1+x)^{-1} \cong 1-x$ for $x \ll 1$ was used and $\Delta P_p \ll <P_p>$ was assumed. With the pump SNR given in Eq. 4.107, the variance of the output phase can be written as

$$<(\vartheta_s - <\vartheta_s>)^2> = \vartheta_{un}^2 \frac{<(\Delta P_p)^2>/<P_p>^2}{(P_{sat}/<P_p> + 2 + <P_p>/P_{sat})^2}$$
$$= \vartheta_{un}^2 \frac{2/\text{SNR}_p}{(P_{sat}/<P_p> + 2 + <P_p>/P_{sat})^2} \equiv <\phi_{xpm}^2> = \sigma_{xpm}^2. \quad (5.16)$$

Thus, the variance of the output phase, that can be interpreted as the pump-induced nonlinear phase noise variance, is inversely proportional to the pump SNR. Although Eq. 5.16 was derived for the (amplified) input signal wave, it is also valid for the generated idler as shown for the pump-induced nonlinear phase noise in the HNLF-based wavelength converters, e.g. in Fig. 4.29. The standard deviation σ_{xpm} calculated with Eq. 5.16 is shown in Fig. 5.12. For a small pump power, $<P_p> \ll P_{sat}$, the nonlinear phase noise variance is proportional to the square of the pump power similar to the result for the HNLF-based wavelength converter given in Eq. 4.113. However, for a high

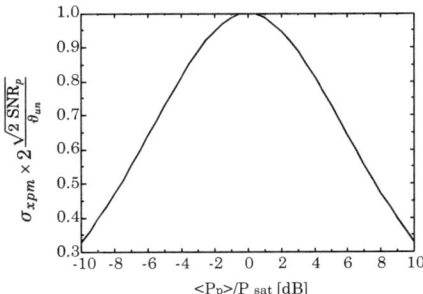

Figure 5.12: Analytically calculated standard deviation of the pump-induced nonlinear phase noise (normalized to its maximum) in the single-pump SOA-based wavelength converter as a function of the pump power (normalized to the SOA saturation power). For the calculation, Eq. 5.16 was used.

pump power $<P_p> \gg P_{sat}$, nonlinear phase noise variance is inversely proportional to $<P_p>^2$. I.e. while the pump-induced nonlinear phase is constantly growing with the pump power in the HNLF-based wavelength converters, a maximum occurs at $P_p = P_{sat}$ in the SOA-based wavelength converters. The reason for this difference is the gain saturation in the SOA.

5.3.2 Pump-Induced Phase Noise: Numerical Results

Due to the complex nonlinear behavior of the SOA, the analytical estimate from the previous section cannot be used for a quantitative determination of the degradation due to nonlinear phase noise in the SOA. For this aim, numerical simulations using the model presented in Sec. 2.3 have to be conducted. The simulation setup is shown in Fig. 5.13a. The pump wave is distorted with AWG noise and injected into the SOA together with the noise-free input signal wave. Still, the exact determination of the nonlinear phase noise variance at the SOA output is difficult because the SOA also produces amplified spontaneous emission noise which is difficult to discriminate from the nonlinear noise. Thus, a third way for the determination of the nonlinear phase noise variance is chosen. The simulation setup is shown in Fig. 5.13b. Instead of distorting the pump amplitude with AWG noise including all frequency components at the same time, the pump wave is sinusoidally amplitude modulated, i.e. with a single frequency component,

$$A_p = \left(1 + m_p \cos(2\pi f_{\sin}t)\right) <A_p>, \qquad (5.17)$$

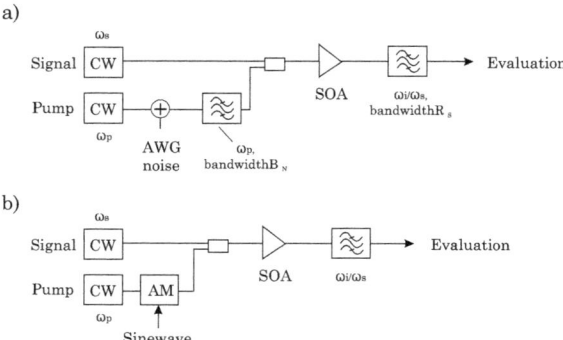

Figure 5.13: a) Simulation setup to characterize pump-induced nonlinear noise by intentionally adding AWG noise to the pump, b) equivalent setup to characterize pump-induced nonlinear noise by intentionally modulating the pump amplitude with a frequency tunable sine wave.

with the pump amplitude modulation index[2] $m_p \ll 1$. The pump power is approximately given by

$$P_p = |A_p|^2 \cong <P_p> + \underbrace{2m_p <P_p> \cos(2\pi f_{\sin} t)}_{\Delta P_{p,\sin}} \quad (5.18)$$

Due to the (sinusoidal) pump power fluctuation $\Delta P_{p,\sin}$, the idler will be phase modulated by XPM which is discussed in more detail in section 2.1.3. Since Eq. 5.15 shows that the idler phase is proportional to any pump power fluctuation, the idler phase modulation will be sinusoidally with $f_{\sin}$. Then, at the SOA output, the idler complex envelope will be proportional to

$$A_i \propto \exp\left(i\beta_i \cos(2\pi f_{\sin} t + \xi_\beta)\right). \quad (5.19)$$

Here, β_i is the idler phase modulation index and ξ_β represents any unknown phase shift. Because β_i can be easily determined at the SOA output a noise transfer function can be constructed by changing the frequency $f_{\sin}$ over different simulations that allows to calculate semi-analytically the transfer of arbitrary noise spectra. In a last step, this procedure is verified by comparing with simulations using the setup shown in Fig. 5.13a.

Fig. 5.14 shows the idler phase modulation index β_i that occurs due to pump

---

[2]Please do not mix this sinusoidal amplitude modulation of the pump in this section with the sinusoidal phase modulation of the pump used to suppress stimulated Brillouin scattering in HNLF-based wavelength converters as discussed in section 4.1.4. While the latter is physical reality, the former is just a simulation technique to determine a transfer function for the pump amplitude noise.

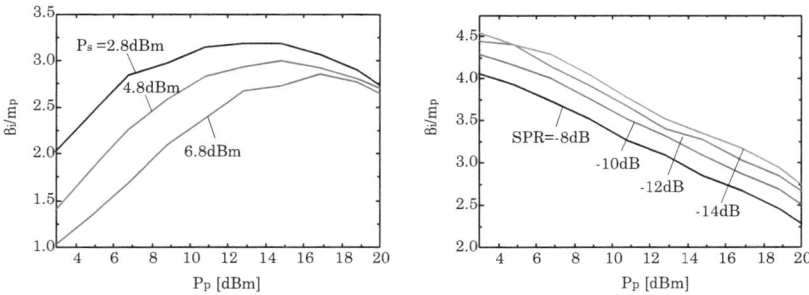

Figure 5.14: a) Idler phase modulation index β_i normalized to the pump amplitude modulation index m_p as a function of the pump power and for different input signal powers, b) same as a) but for different SPR ($f_{\sin}$ = 1.25 GHz, m_p = 0.1). The used parameters are given in the text.

XPM using a sinusoidal amplitude modulation with a frequency of 1.25 GHz. It is normalized to the pump amplitude modulation index and is given as a function of the pump power. The used parameters were L = 1 mm, I_B = 190 mA and $\Delta\lambda$ = −2.5 nm. λ_p was set to the ASE spectral peak. In Fig. 5.14a, the signal power was kept constant. Due to the large alpha factor $\alpha_{H,\mathrm{CDP}}$ = 5 (as listed in section G), the idler phase modulation index is larger by factor 3 at maximum than the pump amplitude modulation index. Thus, a small pump amplitude distortion results in a strong idler phase modulaton. As predicted by the simple analytical model discussed in the previous section, the idler modulation index first increases at low pump powers, then reaches a distinct maximum and falls off after. At low pump powers, the SOA is not saturated and the idler phase distortion increases because the XPM efficiency is proportional to the pump power. At high powers, the idler phase distortion decreases because the SOA is strongly saturated so that the gain and therefore the XPM efficiency is decreasing. For higher signal powers, the phase distortion decreases because the signal also contributes to the SOA saturation. Fig. 5.14b shows the same data but keeping the signal-to-pump power ratio (SPR) constant instead of the signal power. In this case, the idler phase distortion decreases monotonically with increasing pump power due to the SOA saturation. The relative contribution of the signal to the SOA saturation is constant due to the constant SPR. For lower SPR values, the signal contributes stronger to the SOA saturation leading to a decreased XPM efficiency. The transfer function of the process is shown in Fig. 5.15. The used parameters were L = 1 mm, I_B = 190 mA, P_p = 12.8 dBm, P_s = 2.8 dBm and $\Delta\lambda$ = −2.5 nm. λ_p was set to the ASE spectral peak. Due to the limited time

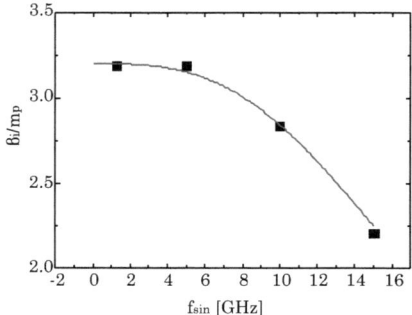

Figure 5.15: Idler phase modulation index β_i normalized to the pump amplitude modulation index m_p as a function of the sine frequency $f_{mathrmsin}$ (square symbols: simulation, straight line: fit, $m_p = 0.1$). The used parameters are given in the text.

constant of the CDP gain contribution, the XPM efficiency falls off quickly beyond 10 GHz. This bandwidth is similar to that of the CDP contribution to the FWM conversion efficiency schematically depicted in Fig. 5.4b (as well as similar to the XGM bandwidth of the SOA). The simulated values were fitted by the third-order low-pass filter function given by

$$H_{npn,p} = \frac{\beta_i(f_{\sin})}{m_p} \cong \frac{\beta_i(f=0)/m_p}{1+(f/f_g)^3} \qquad (5.20)$$

with $\beta_i(f=0)/m_p = 3.2$ and a critical frequency $f_g = 20$ GHz.

As was seen in the previous sections, increasing the length of the SOA is advantageous in terms of conversion efficiency and output idler OSNR. Fig. 5.16 shows the idler phase modulation due to XPM in presence of the sinusoidally amplitude modulated pump for different SOA lengths. The used simulation parameters were $I_B = 190$ mA/mm and $\Delta\lambda = -2.5$ nm. λ_p was set at the ASE spectral peak which changes for the different lengths as given by Tab. G.2. The qualitative behaviour is the same as discussed for Fig. 5.14. However, the idler phase distortion is increasing with the SOA length potentially counteracting the advantages of the long SOA.

5.4 Impact of Signal-Induced Phase Noise

Not only the pump wave, but also the input signal wave can be distorted by noise. In particular, if the AOWC is used within a transmission system, it cannot be avoided that the input signal may be at a low SNR level. In this case, the input signal amplitude noise will lead to nonlinear phase noise due to

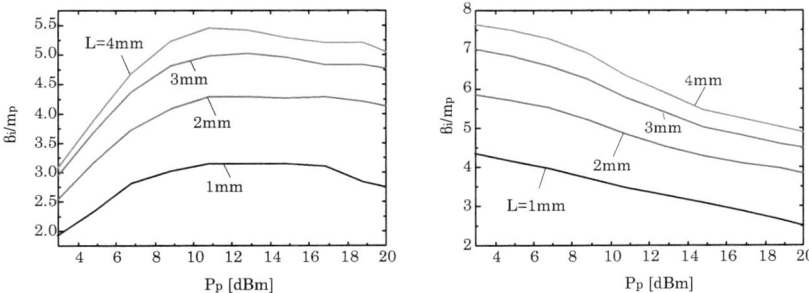

Figure 5.16: a) Idler phase modulation index β_i normalized to the pump amplitude modulation index m_p as a function of the pump power for P_s = 2.8 dBm and for different SOA lengths , b) same as a) but for for SPR = -10 dB (f_{sin} = 1.25GHz, m_p = 0.1). The used parameters are given in the text.

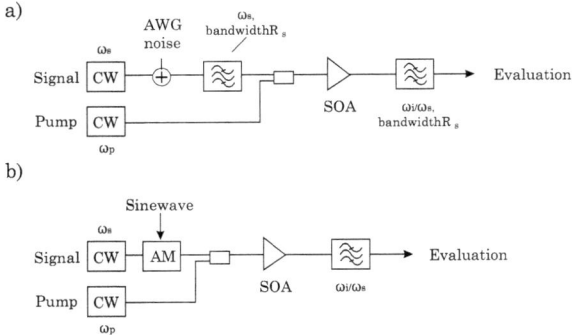

Figure 5.17: a) Simulation setup to characterize signal-induced phase noise by intentionally adding AWG noise to the signal, b) equivalent setup to characterize signal-induced phase noise by modulating the signal amplitude with a frequency-tunable sine wave

self-phase modulation (SPM). The physical origin of this process is discussed in section 2.1.3. The analysis for the HNLF-based wavelength converters in section 4.5 showed that the amount of generated nonlinear phase noise decreases with a lower signal-to-pump power ratio (SPR), i.e. the input signal power should be chosen much lower than the pump power. However, Fig. 5.9b shows that, for the SOA-based single-pump wavelength converter, the SPR cannot be chosen arbitrarily low in order to keep the output OSNR at a high value. Counteracting the output OSNR decrease by increasing the pump power is also limited by the waveguide input power limits. Thus, only a compromise between linear and nonlinear noise performance is possible.

For the simulation of signal-induced nonlinear phase noise, a similar approach was chosen as for the pump-induced nonlinear phase noise in the previous section. Instead of distorting the input signal by AWG noise as shown in Fig. 5.17a, the signal-induced noise is modeled using a sinusoidal amplitude modulation of the input signal amplitude as shown in Fig. 5.17b. By varying the modulation frequency f_{sin}, a noise transfer function is constructed. The input signal shall be given by

$$A_s \propto (1 + m_s \cos(2\pi f_{\text{sin}} t)), \tag{5.21}$$

with the input signal amplitude modulation index m_s. Similar to what was discussed in the previous section, the generated idler will show a sinusoidal phase modulation due to SPM with the same modulation frequency f_{sin},

$$A_i \propto \exp\left(i\beta_i \cos(2\pi f_{\text{sin}} t + \xi_\beta)\right) \tag{5.22}$$

where β_i is the phase modulation index that has to be determined by the simulations and ξ_β is an unknown phase shift. Fig. 5.18 shows the resulting phase modulation index of the idler β_i (normalized to the input signal amplitude modulation index m_s) as a function of the pump power. The pump wave was noise-free in this case. For comparison, the phase modulation index of the idler β_i resulting from a pump amplitude modulation with $m_p = m_s$ is shown. In this case, the signal was kept noise-free. The used parameters were L = 1 mm, I_B = 190 mA and $\Delta\lambda$ = −2.5 nm. λ_p was set to the ASE spectral peak. In Fig. 5.18a, the signal input power is kept constant. For low pump powers, the idler phase modulation due to the input signal wave and the pump wave have similar magnitudes. This is because the signal input power and the pump power are similar. With increasing pump power (and therefore larger SPR), the idler phase modulation due to the pump increases while the phase modulation due to the signal decreases. Fig. 5.18b, shows the same data but for a

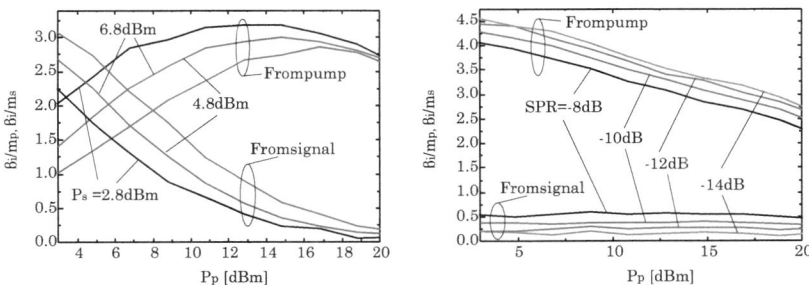

Figure 5.18: a) Normalized idler phase modulation index due to signal amplitude modulation β_i/m_s ($m_s = 0.1$, $f_{sin} = 3.13$ GHz) and normalized idler phase modulation index due pump amplitude modulation β_i/m_p ($f_{sin} = 1.25$ GHz, $m_p = 0.1$), both as a function of the pump power and different input signal powers, b) same as a) but for different signal-to-pump power ratios

constant SPR. Here, it can be clearly seen that the signal-induced idler phase modulation is independent on the pump power. It decreases with decreasing SPR. Thus, a high pump power together with a low SPR is the preferred operation point for the single-pump SOA-based AOWC in order to keep the nonlinear noise as low as possible. This is the same conclusion as for the HNLF-based wavelength converters drawn in section 4.5. Due to the similar physical origin of XPM and SPM, the noise transfer function $H_{npn,s}$,

$$H_{npn,s} = \frac{\beta_i(f_{sin})}{m_s} \cong \frac{\beta_i(f=0)/m_s}{1+(f/f_g)^3}, \tag{5.23}$$

is similar to $H_{npn,p}$ given in Eq. 5.20, in particular with the same critical frequency f_g. Fig. 5.19 shows the dependence of the signal-induced idler phase modulation on the length of the SOA. The used simulation parameters were $I_B = 190$ mA/mm and $\Delta\lambda = -2.5$ nm. λ_p was set at the ASE spectral peak which changes for the different lengths as given by Tab. G.2. In difference to the pump-induced idler phase modulation, the dependence is weak in particular for high pump powers. Thus, an increased SOA length does not yield an increased signal-induced idler phase modulation.

5.5 (O)SNR Penalty due to Pump- and Signal-Induced Phase Noise

In order to calculate the (O)SNR penalty due to the pump- and signal-induced phase noise using Eqs. 3.32 and 3.34, the noise variance of the nonlinear

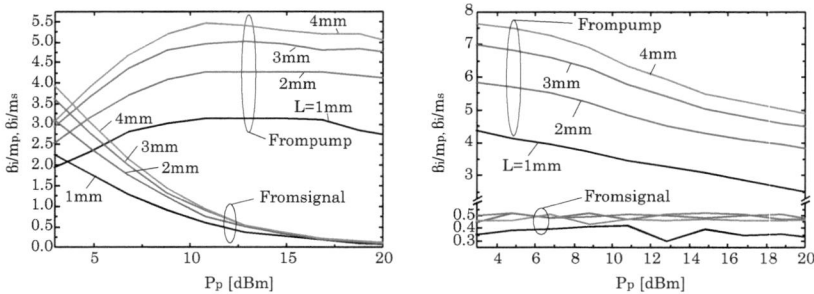

Figure 5.19: a) Normalized idler phase modulation index due to signal amplitude modulation β_i/m_s ($m_s = 0.1$, $f_{sin} = 3.13$ GHz) and normalized idler phase modulation index due pump amplitude modulation β_i/m_p ($f_{sin} = 1.25$ GHz, $m_p = 0.1$), both for a constant input signal power $P_s = 2.8$ dBm and as a function of the pump power and different SOA lengths , b) same as a) but for a constant signal-to-pump power ratio SPR = −10 dB

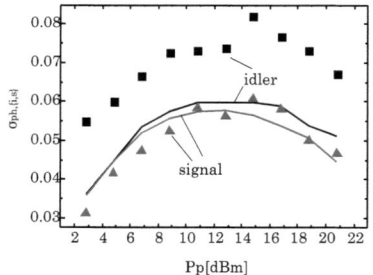

Figure 5.20: Signal and idler phase noise standard deviation as a function of the pump power (symbols: full simulation with noise, straight lines: semianalytical results based on simulation with sinusoidal pump amplitude modulation)

phase noise has to be known. As explained in section 5.3.2, the noise transfer functions $H_{npn,p}(f) = \beta_i(f)/m_p$ and $H_{npn,s}(f) = \beta_i(f)/m_s$ given in Eqs. 5.20 and 5.23 can be used for this aim. Then, the nonlinear phase noise variance is given by

$$\sigma_{npn}^2 = \int_0^{B_N/2} (H_{npn,p})^2 \frac{\rho_{\text{AWG},p}}{<P_p>} df + \int_0^{B_N/2} (H_{npn,s})^2 \frac{\rho_{\text{AWG},s}}{<P_s>} df \qquad (5.24)$$

with the pump noise bandwidth $B_N < R_s$ after Eq. 4.110. The normalized amplitude noise power spectral densities $\rho_{\text{AWG},p}/<P_p>$ and $\rho_{\text{AWG},s}/<P_s>$ are constant because the pump and the input signal wave shall be distorted by AWG noise. Thus, Eq. 5.24 can be written as

$$\sigma_{npn}^2 = \frac{\rho_{\text{AWG},p}}{<P_p>} \int_0^{B_N/2} (H_{npn,p})^2 df + \frac{\rho_{\text{AWG},s}}{<P_s>} \int_0^{B_N/2} (H_{npn,s})^2 df. \qquad (5.25)$$

On the other hand, using Eq. 4.107,

$$<(\Re\{\Delta A_p\})^2> = \int_0^{B_N/2} \rho_{\text{AWG},p} df = \rho_{\text{AWG},p} B_N/2 = \frac{<P_p>}{2\text{SNR}_p}$$

$$<(\Re\{\Delta A_s\})^2> = \int_0^{B_N/2} \rho_{\text{AWG},s} df = \rho_{\text{AWG},s} B_N/2 = \frac{<P_s>}{2\text{SNR}_s}. \qquad (5.26)$$

With this, Eq. 5.25 takes its final form,

$$\sigma_{npn}^2 = \frac{1}{B_N \text{SNR}_p} \int_0^{B_N/2} \left(\frac{\beta_i}{m_p}\right)^2 df + \frac{1}{B_N \text{SNR}_s} \int_0^{B_N/2} \left(\frac{\beta_i}{m_s}\right)^2 df. \qquad (5.27)$$

Fig. 5.21 shows the signal and idler phase standard deviations calculated using Eq. 5.27. The normalized idler phase modulation index β_i/m_p results from a simulation determining the idler phase modulation index resulting from a pump amplitude modulation for different modulation frequencies and pump power. These idler phase standard deviations are compared to results of full numerical simulations with 2^{18} samples using the setup shown in Fig. 5.14, i.e. with a pump wave distorted by AWG noise. The used parameters were L = 1 mm, I_B = 190 mA and $\Delta\lambda$ = −2.5 nm, P_s = 2.8 dBm, SNR_p = 30 dB, B_N = 40 GHz. λ_p was set to the ASE spectral peak. Additionally to the idler phase standard deviation, also the phase standard deviation of the amplified output signal is shown. Both values should be identical (compare to the HNLF-based wavelength converters, Fig. 4.29). The difference shown in Fig. 5.14 results from a larger ASE noise contribution added by the SOA to the idler than to the amplified signal. The low conversion efficiency (compare to Fig. 5.5) results in a lower idler output power, i.e. in a lower output OSNR, which in turn increases the phase noise standard deviation. Thus, a comparison of semianalytical results obtained by Eq. 5.27 to the phase standard deviation of amplified output signal is more reliable because this standard deviation is indeed dominated by the nonlinear phase noise. The comparison yields a good match validating the approach using the sine modulation of the pump used in the previous sections.

Now, the (O)SNR penalties can be calculated using the simulated idler phase modulation index β_i from the previous sections, Eq. 5.27 and Eqs. 3.32 and 3.34. Fig. 5.21a shows the (O)SNR penalty at a BER of 10^{-4} resulting from pump-induced nonlinear phase noise as a function of the pump power. Two different pump SNR values and two different modulation formats, directly detected DQPSK and coherently detected 8-PSK, are considered. The parameters for the determination of β_i were L = 1 mm, I_B = 190 mA, $\Delta\lambda = -2.5 nm$ and SPR = -14 dB, the used β_i is shown in Fig. 5.14. For simplicity, $H_{npn,p}(f) \cong H_{npn,p}(0)$ was set which is approximately valid for noise bandwidths $B_N \leq 25$ GHz when using f_g = 20 GHz as characterized for the 1-mm long SOA in section 5.3.2. Signal-induced phase noise was not taken into account. The graph shows that a pump SNR value of > 40 dB is needed to avoid significant (O)SNR penalties for a single conversion. In Fig. 5.21b, the (O)SNR penalty at a BER of 10^{-4} resulting from signal-induced nonlinear phase noise is shown as a function of the input signal SNR. Here, the same parameter settings as in Fig. 5.21a were assumed except of the different SPR values while the pump-induced nonlinear phase noise was not taken into account. The used β_i is shown in Fig. 5.18 and is independent on the pump power. To avoid significant penalties, the input signal SNR must be > 25 dB or the SPR must be chosen to < - 14 dB. For orientation, the required SNR values to reach the BER of 10^{-4} (as shown in Fig. 3.4) are also marked. Fig. 5.22 shows the dependency of the (O)SNR penalties on the SOA length. The used values for β_i are shown in Fig. 5.19b. Other used parameters were I_B = 190 mA/mm and $\Delta\lambda = -2.5$ nm. Similarly to the conclusions drawn in the previous sections, the graph shows that an increasing SOA length increases the requirements on the pump SNR while the tolerance to signal noise decreases from 1 mm to 2 mm length but does not decrease further.

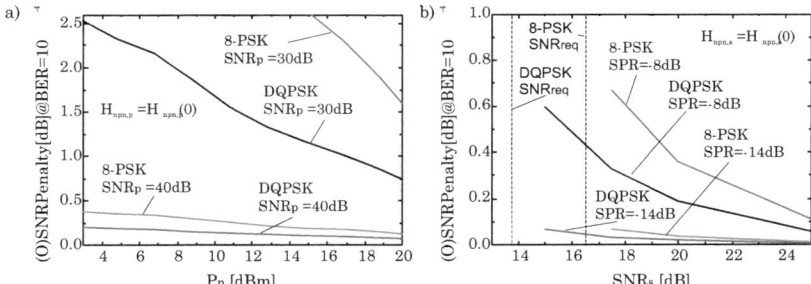

Figure 5.21: a) (O)SNR penalty at a BER = 10^{-4} for DQPSK and 8-PSK as a function of the pump power and for different pump SNR values (SPR = -14 dB), b) (O)SNR penalty at a BER = 10^{-4} for DQPSK and 8-PSK as a function of the signal SNR and for different SPR values (P_p = 12.8 dBm)

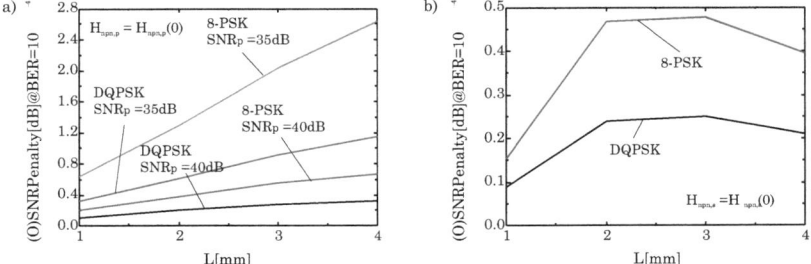

Figure 5.22: a) (O)SNR penalty at a BER = 10^{-4} for DQPSK and 8-PSK as a function of the SOA length and for different pump SNR values (SPR = -14 dB, P_p = 12.8 dBm), b) (O)SNR penalty at a BER = 10^{-4} for DQPSK and 8-PSK as a function of the SOA length and for a signal SNR of 20 dB (SPR = -14 dB). The other parameters are given in the text.

Chapter 6
Conclusions

In this thesis, transparent parametric amplifier and wavelength converters have been theoretically investigated regarding their capability to deal with higher-order phase-modulated signals. This format transparency is one of the key features for a practical component finding applications in future optical networks. The analytical and numerical investigations concentrated on the identification of phase distortions, the evaluation of their impact in terms of BER and their mitigation or compensation. Two different components were considered, highly nonlinear fibers (HNLF) and semiconductor optical amplifiers (SOA).

Parametric amplification and wavelength conversion in HNLF

For the FWM-based FOPAs, there are three different options given by the three different FWM processes in the HNLF. This is single-pump FWM, phase-conjugating dual-pump FWM and non-phase conjugating (only frequency-converting) dual-pump FWM. All three processes provide full bitrate and modulation format transparency if pumped with continuous wave (CW) signals. Furthermore, they are strongly polarization-dependent, but, due to the linearity of the wavelength conversion, diversity schemes can be applied. The single-pump and phase-conjugation process both provide parametric amplification and phase-conjugating wavelength conversion with limited tunability, i.e. for a given input signal wavelength, the output signal wavelength is fixed. For high conversion efficiencies > 0 dB, the pumps have to be phase-modulated to suppress stimulating Brillouin scattering. In this operation regime, the noise figure closely approaches the quantum of 3 dB and flat gain / conversion spectra with bandwidths > 50 nm can be obtained. Both processes can be used for fiber optic parametric amplifiers and for Kerr compensation using midspan spectral inversion. By contrast, the frequency-conversion process does not provide parametric amplification but non-phase conjugating and fully tunable wavelength conversion. The maximum conversion efficiency is 0 dB as well as the ideal noise figure. This process is ideal for contention resolution and all-optical routing in combination with an arrayed waveguide.

When used as a wavelength converter, the major phase distortion results from the transfer of the pump-phase modulation to the converted signal. This leads to high (O)SNR penalties that grow with the degree of the phase modulation format effectively degrading the format transparency as experimentally demonstrated in [42]. Two different compensation schemes have been investigated. The first is co- and counterphasing of the pumps which can be applied to the phase-conjugation and the frequency-conversion process, respectively. Although the tolerances are critical, this scheme provides nearly ideal compensation of phase distortion which could be verified also in system experiments converting 80 Gb/s DQPSK signals with a conversion gain of 15 dB [109]. The second scheme is applicable to all three conversion processes but is restricted to coherently detected formats because it relies on electronic signal processing. It was verified in experiments using 20 Gb/s QPSK with a conversion gain of -3 dB [37] that also this scheme leads to a full compensation of the phase distortions. However, the highest modulation frequency that can be compensated depends on the symbol rate.

Another issue for all three processes is the increase of the carrier linewidth due to the wavelength conversion process because the laser phase noise of the pump wave adds to that of the input signal. Since directly detected formats are relatively robust against laser phase noise, this is in particular an issue for coherently detected formats. Here, the required average linewidth per laser (including the LO laser at the receiver) halves for a single conversion and decreases further for cascaded wavelength conversions. To avoid this, pump waves with laser linewidths much smaller than 100 kHz have to be used for symbol rates around 50 GBd. For the FC FWM process the linewidth increase can be canceled out by phase locking of the two pumps as was experimentally shown for 10 Gb/s QPSK in [105].

When operating SP FWM and PC FWM devices as parametric amplifiers, neither pump-phase modulation nor the pump laser linewidth is an issue. However, pump-induced nonlinear phase noise is a problem in particular in the high gain regime and for cascaded operation. A solution is the use of pump waves with relative intensity noise levels below -160 dB/Hz. A comparison to the impact of nonlinear amplitude noise on 16-QAM signals shows that nonlinear phase noise dominates the degradation. This was experimentally demonstrated for parametric amplification of 28-Gbd NRZ-16QAM signals [131]. Finally, to avoid nonlinear phase noise generated by self-phase modulation in the parametric devices, the signal output power should be more than 10 dB below the pump power, or, equivalently, pump depletion must be avoided.

Parametric wavelength conversion in SOA

For parametric wavelength conversion in SOA, also single pump and dual pump options exists. However, only the single-pump scheme was investigated since no principal advantage was expected from the dual-pump schemes. In terms of conversion efficiency, conversion bandwidth and noise figure, the SOA-based converter performs significantly worse than the HNLF-based converters. For a 1 mm long SOA biased at 190 mA, the 3-dB conversion bandwidth is only about 0.1 nm. For a wavelength detuning of about 1 nm between pump and input signal wave, the conversion efficiency has dropped to values below -10 dB while the noise figure easily exceeds 15 dB (excluding coupling losses) due to the inherent generation of ASE. Increasing the SOA length to about 4 mm is advantageous both increasing the conversion efficiency and decreasing the noise figure by about 10 dB.

In terms of phase distortions, the SOA-based converter also suffers from the linewidth increase due to the wavelength conversion while a pump-phase modulation is not necessary and the related degradation do not occur. Similar to the HNLF-based amplifiers and converters, pump XPM and signal SPM generate nonlinear phase noise. Since the signal input power cannot be chosen arbitrarily low because of the ASE noise floor, a compromise between linear noise performance and the generation of nonlinear noise has to be taken. This explains experimental results where a relatively strong degradation for DQPSK signals was observed [140, 143]. The best performance is expected for very high pump powers exceeding 15 dBm and relatively low signal powers up to 15 dB below the pump power which is confirmed by experimental characterizations. In this regime, signal SPM is critical only for input signal SNR value below 20 dB. However, pump XPM generated nonlinear noise is at a high level requiring low RIN pump waves < -140 dB/Hz. The RIN tolerance decreases further for long SOAs. Generally, SOAs with a low alpha factor are advantageous to avoid this issue.

Summary and outlook

In summary, the investigations show that single-pump and phase-conjugation based parametric amplifiers and wavelength converters with nearly ideal format transparency and high gain can be realized in HNLF if additional complexity (co- or counterphased pump-phase modulation, low pump laser linewidth, low pump-laser RIN) is accepted. Together with the option of low loss splicing to the SSMF, this makes these devices ideal candidates for broadband multi-wavelength devices in the transmission link providing parametric amplification, Kerr compensation by optical phase conjugation, waveband monitoring or waveband conversion to longer wavelengths in order to take profit from new transmission fibers. Furthermore, also the use for regeneration and further optical signal processing can be predicted although the question arises if HNLF-based converters are too bulky for single wavelength devices of which many are needed at a network node or in a transmission link. The same issue comes into mind when proposing frequency-conversion FWM based parametric wavelength converters for contention resolution or for routing in burst and packet switching nodes despite the excellent performance in terms of phase distortions, noise figure and wavelength tunability.

For the latter mentioned network functions, integrable solutions like SOA-based wavelength converters seem to be the natural solution despite the significantly worse performance in comparison to the HNLF-based converters. However, integrable passive devices like silicon nanowires [144] are a strong competitor due to the better noise performance, but CW-pumped parametric amplification is still to be shown. But also SOA technology moves forward so that a new class of high power, low noise SOAs with 0.8 W output power and 5.5 dB fiber coupled noise figure was recently demonstrated [145, 146]. A key advantage for the SOA-based converter arrays could be the fact that they can be integrated together with the pump laser array.

A review of recent experiments with PPLN based $\chi^{(2)}$ parametric wavelength converters [147, 148] confirms that those device do not suffer from additional phase distortions except of the increase of the laser phase noise. The absence of SPM and XPM makes $\chi^{(2)}$ devices in principle superior wavelength converters for phase modulated signals. However, in concurrence to the HNLF-based converters, the worse noise figure due to high coupling losses to SSMF and low conversion efficiencies < 0 dB is a strong disadvantage (although CW-pumped parametric amplification was very recently demonstrated [38, 39]), also viewing the fact that the phase distortions in the HNLF can be avoided by some additional effort as shown above. Because a PPLN device is much more bulky in comparison to a SOA or to a silicon waveguide and needs hybrid integration on common InP or Si platforms, also the use in wavelength converter arrays seems to be questionable despite its superior performance.

In the present thesis, only single channel effects in the wavelength converters have been investigated. Furthermore, also no polarization effects have been considered. Thus, further investigations on polarization-insensitive parametric amplification and wavelength conversion [149] of signals consisting of many WDM signals in realistic transmission systems should be conducted [150]. Here, interchannel FWM and XPM and cross-gain modulation due to pump depletion are additional effects to be considered. The realization of practical phase-sensitive parametric amplifiers and wavelength converters is another interesting research topic [20, 22]. In particular, the question arises in which way these amplifiers can be used for broadband amplification [13].

The advent of coherent detection in optical transmission systems enables the use of electronic signal processing for impairment mitigation. This was used within this thesis to compensate for phase distortions due to the pump-phase modulation in HNLF-based parametric wavelength converters. Recently, algorithms to compensate for amplitude and phase distortions in SOAs used as amplifiers have been proposed relying on nonlinear back propagation [151, 152, 153]. These two examples illustrate the great possibilities that are provided by the combination of optical and electronic signal processing.

Finally, as mentioned above, the performance of parametric amplifiers and wavelength converters depends crucially on the quality of the pump laser. Thus, the lack in availability of high-power spectrally narrow single-mode laser diodes at Watt-level were one of the major obstacles preventing the deployment these devices. While appropriate lasers with half a Watt output power were recently reported [154], further developments in this field are a remaining task.

Appendix A

Definition of the Fourier Transforms

Throughout this thesis, the continuous Fourier transform shall be defined by

$$\mathcal{F}[A(t)] = \tilde{A}(\omega) = \int_{-\infty}^{\infty} A(t)e^{i\omega t}dt. \tag{A.1}$$

Likely, the inverse continuous Fourier transform is defined by

$$\mathcal{F}^{-1}[\tilde{A}(\omega)] = A(t) = \frac{1}{2\pi}\int_{-\infty}^{\infty} \tilde{A}(\omega)e^{-i\omega t}d\omega. \tag{A.2}$$

The time-discrete Fourier transform is defined by

$$\mathcal{F}[A_n] = \tilde{A}(\Omega) = \sum_{k=-\infty}^{\infty} A_k e^{ik\Omega}. \tag{A.3}$$

Thereby, $\Omega = 2\pi\omega/\omega_T$. ω_T is the sampling frequency. The inverse time-discrete Fourier transform is given by

$$\mathcal{F}^{-1}[\tilde{A}(\Omega)] = A_n = \frac{1}{2\pi}\int_{-\pi}^{\pi} \tilde{A}(\Omega)e^{-in\Omega}. \tag{A.4}$$

Appendix B

Derivation of the Nonlinear Wave Equation

This derivation follows [50, p. 25]. In their differential form, Maxwell's equations are given by

$$\nabla \times \vec{H} = \frac{\partial \vec{D}}{\partial t} + \vec{J} \tag{B.1}$$

$$\nabla \times \vec{E} = -\frac{\partial \vec{B}}{\partial t} \tag{B.2}$$

$$\nabla \cdot \vec{D} = \rho \tag{B.3}$$

$$\nabla \cdot \vec{B} = 0 \tag{B.4}$$

$$\tag{B.5}$$

with the electric field vector $\vec{E}$, the magnetic field vector $\vec{H}$, the electric flux density vector $\vec{D}$, the magnetic flux density vector $\vec{B}$, the current density vector $\vec{J}$ and the free charge density ρ. Within the field of nonlinear optics, it can be typically assumed that the material does contain no free charges and no free currents, so that

$$\rho = 0 \tag{B.6}$$

and

$$\vec{J} = \sigma \vec{E} = 0. \tag{B.7}$$

Furthermore, the material is assumed to be nonmagnetic, so that

$$\vec{B} = \mu_0 \vec{H}. \tag{B.8}$$

with the vacuum permeability μ_0. However, the material is allowed to be nonlinear in the sense that

$$\vec{D} = \epsilon_0 \vec{E} + \vec{P} \tag{B.9}$$

where ϵ_0 denotes the vacuum permittivity and, in general, the material polarization vector $\vec{P}$ depends nonlinearly on the local value of the electric field vector $\vec{E}$.

Now, the nonlinear wave equation is derived. Taking the curl of Eq. (B.2), replacing $\vec{B}$ by $\vec{H}$ through Eq. (B.8) and interchanging the space and time derivatives on the right-hand side gives

$$\nabla \times \nabla \times \vec{E} = -\mu_0 \frac{\partial}{\partial t} \nabla \times \vec{H}. \tag{B.10}$$

Inserting Eqs (B.1) together with Eq. (B.7) leads to

$$\nabla \times \nabla \times \vec{E} = -\mu_0 \frac{\partial^2 \vec{D}}{\partial t^2}. \tag{B.11}$$

With the electric material equation (B.9), it follows that

$$\nabla \times \nabla \times \vec{E} + \frac{1}{c_0^2} \frac{\partial^2 \vec{E}}{\partial t^2} = -\frac{1}{\epsilon_0 c_0^2} \frac{\partial^2 \vec{P}}{\partial t^2}. \tag{B.12}$$

with the velocity of light in vacuum given by $c_0 = (\epsilon_0 \mu_0)^{-\frac{1}{2}}$. Usually, Eq. B.12 is simplified using the identity

$$\nabla \times \nabla \times \vec{E} = \nabla \left(\nabla \cdot \vec{E} \right) - \Delta \vec{E}. \tag{B.13}$$

and neglecting the first term on the right-hand side. This term vanishes for linear source-free isotropic media since then Eq. (B.3) implies $\nabla \cdot \vec{E} = 0$. In nonlinear optics, it is generally nonvanishing but it can be dropped for most cases of interest [49, p. 71]. Then, the nonlinear wave equation is finally given by

$$\Delta \vec{E} - \frac{1}{c_0^2} \frac{\partial^2 \vec{E}}{\partial t^2} = \frac{1}{\epsilon_0 c_0^2} \frac{\partial^2 \vec{P}}{\partial t^2}. \tag{B.14}$$

Appendix C

Perturbation Theory

In this appendix, it is shown how to solve Eq. 2.15 using perturbation theory as it was done in [51, p. 40] and [50, p. 34]. Eq. 2.15 is written in the form

$$\left(\Delta_T + \frac{\omega^2}{c_0^2}\epsilon(x,y,\omega) - \zeta\right)F(x,y) = 0 \qquad (C.1)$$

where $\Delta_T = \frac{\partial^2}{\partial x^2} + \frac{\partial^2}{\partial y^2}$ is the transversal Laplacian operator and $\zeta = \tilde{\beta}^2$. Furthermore,

$$\epsilon(x,y,\omega) = \epsilon_b(x,y,\omega) + \delta_p\Delta\epsilon(x,y,\omega) \qquad (C.2)$$

$$\tilde{\beta}(\omega) = \beta(\omega) + \delta_p\Delta\beta(\omega) \qquad (C.3)$$

$$\zeta = \zeta_0 + \delta_p\Delta\zeta \cong \beta(\omega)^2 + \delta_p 2\beta(\omega)\Delta\beta(\omega) \qquad (C.4)$$

$$F(x,y) = F_0(x,y) + \delta_p\Delta F(x,y) \qquad (C.5)$$

where δ_p is the perturbation parameter which is arbitrarily small. Insertion into Eq. C.1 and ordering after powers of δ_p yields

$$\delta_p^0 : \left(\Delta_T + \frac{\omega^2}{c_0^2}\epsilon_b(\omega) - \zeta_0\right)F_0 = 0 \qquad (C.6)$$

$$\delta_p^1 : \Delta_T\Delta F + \left(\frac{\omega^2}{c_0^2}\epsilon_b(\omega) - \zeta_0\right)\Delta F + \left(\frac{\omega^2}{c_0^2}\Delta\epsilon - \Delta\zeta\right)F_0 = 0. \qquad (C.7)$$

Now, Eq. C.7 is multiplied with F_0^* and integrated over the whole x-y plane,

$$\iint_{-\infty}^{\infty} F_0^*\Delta_T\Delta F\, dxdy + \iint_{-\infty}^{\infty}\left(\frac{\omega^2}{c_0^2}\epsilon_b(\omega) - \zeta_0\right)F_0^*\Delta F\, dxdy + \iint_{-\infty}^{\infty}\frac{\omega^2}{c_0^2}\Delta\epsilon(\omega)|F_0|^2 dxdy$$

$$= \Delta\zeta\iint_{-\infty}^{\infty}|F_0|^2 dxdy. \qquad (C.8)$$

Using the second identity of Green gives

$$\iint_{-\infty}^{\infty}F_0^*\Delta_T\Delta F\, dxdy = \iint_{-\infty}^{\infty}\Delta F\Delta_T F_0^*\, dxdy + \underbrace{\int_{\partial A}\left(F_0^*\frac{\partial\Delta F}{\partial n} - \Delta F\frac{\partial F_0^*}{\partial n}\right)dS}_{=0} \qquad (C.9)$$

where $\partial/\partial n$ the normal derivative with respect to the integration surface. The last term disappears because F_0 and ΔF as well as their normal derivatives disappear for $x, y \to \infty$. Insertion of Eq. C.9 in Eq. C.8 yields

$$\iint_{-\infty}^{\infty} \Delta F \underbrace{\left(\Delta_T + \frac{\omega^2}{c_0^2}\epsilon_b(\omega) - \zeta\right)F_0^*}_{=0} dxdy + \iint_{-\infty}^{\infty} \frac{\omega^2}{c_0^2}\Delta\epsilon(\omega)|F_0|^2 dxdy = \Delta\zeta \iint_{-\infty}^{\infty} |F_0|^2 dxdy$$

where Eq. C.6 was used. Solving for $\Delta\zeta \cong 2\beta\Delta\beta$ and $F_0 \to F$ yields Eq. 2.17.

Appendix D

Dispersion Characteristics

In this appendix, the relationship between the dispersion coefficients 2.20 and the experimentally measurable dispersion $\underline{D}$ is shown. From a theoretical point of view, the dispersion characteristics of the HNLF is given by the propagation constant β. For convenience, it is usually expanded into the Taylor series given by Eq. 2.19,

$$\beta(\omega) = \sum_{n=0}^{4} \frac{\beta_n}{n!}(\omega - \omega_0)^n, \tag{D.1}$$

with the coefficients given in Eq. 2.20

$$\beta_n = \left.\frac{d^n \beta}{d\omega^n}\right|_{\omega=\omega_0}. \tag{D.2}$$

Thereby, β_0 and β_1 represent the inverses of the phase and the group velocity at ω_0, respectively. β_2, β_3 and β_4 are called the second, third and fourth order dispersion coefficients.

Experimentally, the dispersion characteristics are often derived from the group delay $\tau = 1/v_{gr}$ which can be measured directly (in contrast to the propagation constant). It is connected to the propagation constant by

$$\tau = \frac{1}{v_{gr}} = \frac{d\beta}{d\omega}. \tag{D.3}$$

A Taylor expansion of τ in the wavelength domain around $\lambda_0 = 2\pi c/\omega_0$ yields

$$\tau = \tau_0 + \underline{D}(\lambda - \lambda_0) + \frac{1}{2}\underline{S}(\lambda - \lambda_0)^2 + \frac{1}{6}\frac{d\underline{S}}{d\lambda}(\lambda - \lambda_0)^3. \tag{D.4}$$

Here, $\underline{D}$ and $\underline{S}$ are commonly called dispersion and dispersion slope, respectively. Using the relation

$$\frac{d}{d\lambda} = -\frac{\omega^2}{2\pi c}\frac{d}{d\omega} \tag{D.5}$$

the coefficients β_2, β_3, β_4 on the one hand and $\underline{D}$, $\underline{S}$ and $d\underline{S}/d\lambda$ on the other hand can be connected by

$$\underline{D} = \left.\frac{d\tau}{d\lambda}\right|_{\lambda=\lambda_0} = -\frac{\omega_0^2}{2\pi c}\beta_2 \tag{D.6}$$

$$\underline{S} = \left.\frac{d^2\tau}{d\lambda^2}\right|_{\lambda=\lambda_0} = \frac{2\omega_0^3}{(2\pi c)^2}\beta_2 + \frac{\omega_0^4}{(2\pi c)^2}\beta_3 \tag{D.7}$$

$$\frac{d\underline{S}}{d\lambda} = \left.\frac{d^3\tau}{d\lambda^3}\right|_{\lambda=\lambda_0} = -\frac{6\omega_0^4}{(2\pi c)^3}\beta_2 - \frac{6\omega_0^5}{(2\pi c)^3}\beta_3 - \frac{\omega_0^6}{(2\pi c)^3}\beta_4. \tag{D.8}$$

Appendix E

Typical HNLF parameters

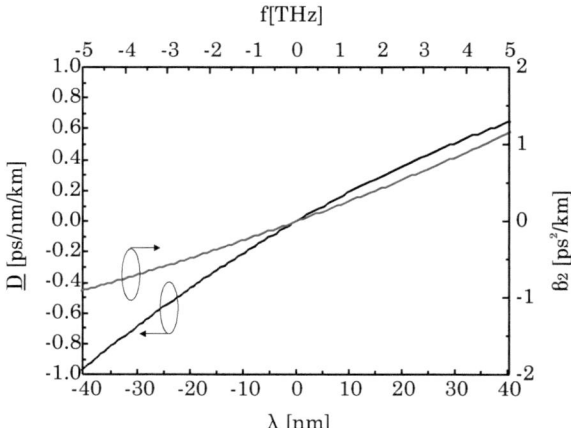

Figure E.1: Dispersion $\underline{D} = d\tau/d\lambda$ as a function of the relative wavelength $\lambda - \lambda_{zd}$ and second-order dispersion coefficient $\beta_2 = d^2\beta/d\omega^2$ as a function of the relative frequency $f - f_{zd}$. Both function were calculated using Eqs. 2.19 and D.4. The parameters are given in Table E.1.

Table E.1: Physical Parameters of SSMF and HNLF

Description	Symbol	SSMF	HNLF
Refractive index of pure silica	n_1	1.45	
Core radius	$d/2$	4 μm	1.5 μm
Relative index step	Δn	0.003	0.03
Nonlinear refractive index	n_2	$2.35 \times 10^{-20} m^2/W$	$3.7 \times 10^{-20} m^2/W$
Absorption coefficient	α_0	0.2 dB/km	0.5 dB/km
2nd order dispersion coefficient 1550 nm	β_2	-27 ps^2/km	0
3rd order dispersion coefficient 1550 nm	β_3	0.132 ps^3/km	0.033 ps^3/km
4th order dispersion coefficient 1550 nm	β_4	-	2.5×10^{-4} ps^4/km
Nonlinear coefficient	γ	1 $(Wkm)^{-1}$	10 $(Wkm)^{-1}$
Effective area	A_{eff}	80 μm^2	12 μm^2
Zero-dispersion wavelength	λ_{zd}	1312 nm	1550 nm
Brillouin frequency shift	Ω_B	10 GHz	
Brillouin peak gain	$g_B(0)$	$3-5 \times 10^{-11} m/W$	
Brillouin bandwidth	$\Delta \nu_B$	20-50 MHz	
Raman frequency shift	Ω_R	13 THz	
Raman peak gain	$g_R(0)$	$1 \times 10^{-13} m/W$	
Raman gain bandwidth	$\Delta \nu_B$	30 THz	
PMD parameter	D_p	0.1-1 $ps/\sqrt{km}$	0.2 $ps/\sqrt{km}$

Appendix F

Calculation of the FIR filter coefficients

In this appendix, the calculation the coefficients of the FIR filter given in 2.85 used in the time-domain model of the SOA is shown as done in [72]. Using Eq. 2.84, the power gain per section is given by

$$|G_l(\omega)|^2 = \exp((\Gamma g(\omega) - a_{\text{int}})\Delta z) \tag{F.1}$$

$$\cong 1 + (\Gamma g(\omega) - a_{\text{int}})\Delta z \tag{F.2}$$

$$= 1 - a_{\text{int}}\Delta z + 3\Gamma g_{p,2}\Delta z \left(\frac{\Delta\omega - \Delta\omega_z}{\Delta\omega_z - \Delta\omega_{p,2}}\right)^2 + 2\Gamma g_{p,3}\Delta z \left(\frac{\Delta\omega - \Delta\omega_z}{\Delta\omega_z - \Delta\omega_{p,3}}\right)^3 \tag{F.3}$$

where $\Delta\omega_x = \omega_x - \omega_0$. The power gain per section provided by the FIR filter is given by Eq. 2.85,

$$|G_{\text{FIR}}(\omega)|^2 = c_{1,l}^2 + |c_{2,l}|^2 + 2c_{1,l}\Re\{c_{2,l}\}\cos(\Delta\omega\Delta t) - 2c_{1,l}\Im\{c_{2,l}\}\sin(\Delta\omega\Delta t) \tag{F.4}$$

$$= c_{1,l}^2 + |c_{2,l}|^2 + 2c_{1,l}\Re\{c_{2,l}\}\left(1 - \frac{\Delta\omega^2\Delta t^2}{2}\right) - 2c_{1,l}\Im\{c_{2,l}\}\Delta\omega\Delta t \tag{F.5}$$

$$= \underbrace{c_{1,l}^2 + |c_{2,l}|^2 + 2c_{1,l}\Re\{c_{2,l}\}}_{=\hat{a}} \underbrace{-2c_{1,l}\Im\{c_{2,l}\}}_{=-\hat{b}}\Delta t\Delta\omega \underbrace{-c_{1,l}\Re\{c_{2,l}\}}_{=-\hat{d}}\Delta t^2\Delta\omega^2 \tag{F.6}$$

Comparing Eqs. F.3 and F.6 yields the coefficients $\hat{a}$, $\hat{b}$ and $\hat{d}$,

$$\hat{a} = 1 - a_{\text{int}}\Delta z + \frac{3\Gamma g_{p,2}\Delta z \Delta\omega_z^2}{(\Delta\omega_z - \Delta\omega_{p,2})^2} - \frac{2\Gamma g_{p,3}\Delta z \Delta\omega_z^3}{(\Delta\omega_z - \Delta\omega_{p,3})^3} \tag{F.7}$$

$$\hat{b} = -\frac{1}{\Delta t}\left(\frac{3\Gamma g_{p,2}\Delta z \Delta\omega_z}{(\Delta\omega_z - \Delta\omega_{p,2})^2} - \frac{3\Gamma g_{p,3}\Delta z \Delta\omega_z^2}{(\Delta\omega_z - \Delta\omega_{p,3})^3}\right) \tag{F.8}$$

$$\hat{d} = \frac{1}{\Delta t^2}\left(\frac{3\Gamma g_{p,2}\Delta z}{(\Delta\omega_z - \Delta\omega_{p,2})^2} - \frac{6\Gamma g_{p,3}\Delta z \Delta\omega_z}{(\Delta\omega_z - \Delta\omega_{p,3})^3}\right). \tag{F.9}$$

Using $\hat{b}^2 + \hat{d}^2 = c_{1,l}^2|c_{2,l}|^2$, $c_{2,l}$ can be eliminated from $\hat{a}$ yielding a bi-quadratic equation for $c_{1,l}$,

$$0 = c_{1,l}^4 - (\hat{a} + 2\hat{d})c_{1,l}^2 + \hat{b}^2 + \hat{d}^2. \tag{F.10}$$

The solution is
$$c_{1,l} = \sqrt{\frac{\hat{a} + 2\hat{d}}{2} + \sqrt{\frac{\hat{a}^2}{4} - \hat{b}^2 + \hat{a}\hat{d}}}. \quad (F.11)$$

$c_{2,l}$ follows from $c_{1,l}$ as
$$c_{2,l} = -\frac{\hat{d}}{c_1} - i\frac{\hat{b}}{c_1}. \quad (F.12)$$

Appendix G

Simulation Parameters for SOA

Table G.1: Simulation Parameters for the SOA

Symbol	Value	Description
Δt	25 E-15 s	Sampling Interval
a_N	3 E-20 m^2	Differential gain
N_{tr}	0.9 E24 m^{-3}	Material Transparency Carrier Density
$\bar{a}$	0.5	Linear Gain Fitting Coefficient
ω_g	1.205E15 Hz	Bandgap Frequency ($\equiv \lambda_g = 1565 nm$)
ω_0	1.207E15 Hz	Reference Frequency ($\equiv \lambda_0 = 1560 nm$)
b_0	2.5 E-11 m^3/s	Peak Frequency Shift Coefficient
ω_{z0}	1.168E15 Hz	Begin of Zero Gain Region ($\equiv \lambda_{z0} = 1615 nm$)
z_0	-1.935E-12 Hz	Zero Gain Frequency Shift Coefficient
ω_c	1.548E15 Hz	Correction of Gain Peak
$\bar{b}$	0.65	Gain Peak Shift Fitting Coefficient
n_G	3.56	Group Index
a_{int}	2250/m	Internal loss
w_w	1.2E-6 m	Active Region Width
d_w	0.2E-6 m	Active Region Height
Γ	0.38	Mode Confinement Factor
Γ_2	1.2	TPA Confinement Factor
A_{nr}	1.25E8 1/s	Unimolecular Non-Radiative Recombination Coefficient
B_{sp}	2.5E-16 m^3/s	Bimolecular Spontaneous Radiative Recombination Coeffcient
C_{Auger}	0.9E-40 m^6/s	Auger recombination Coefficient
τ_{CH}	850 E-15 s	Carrier Heating Time Constant
τ_{SHB}	125 E-15 s	Spectral Hole Burning Time Constant
ϵ_{CH}	0.6 E-23 m^3	Gain Suppression Coefficient (CH)
ϵ_{SHB}	0.5 E-23 m^3	Gain Suppression Coefficient (SHB)
ϵ_{FCA}	0.1 E-24 m^3	Gain Suppression Coefficient (FCA)
ϵ_{TPA}	1.25 E-23 m^3	Gain Suppression Coefficient (TPA)
β_{TPA}	4 E-21 m^2	TPA Coefficient
$\alpha_{H,CH}$	3.5	Linewidth enhancement factor (CH)
$\alpha_{H,SHB}$	0.1	Linewidth enhancement factor (SHB)
$\alpha_{H,FCA}$	0.1	Linewidth enhancement factor (FCA)
$\alpha_{H,TPA}$	-2.25	Linewidth enhancement factor (TPA)
$\alpha_{H,CDP}$	5	Linewidth enhancement factor (CDP)
CL	0	Coupling loss

Table G.2: Simulation Parameters for the SOA

SOA Length	ASE peak wavelength relative to reference wavelength
L	$\lambda_{\text{ASE,peak}} - \lambda_0$
1 mm	-35.5 nm
2 mm	-17.8 nm
3 mm	-10.5 nm
4 mm	-7.3 nm

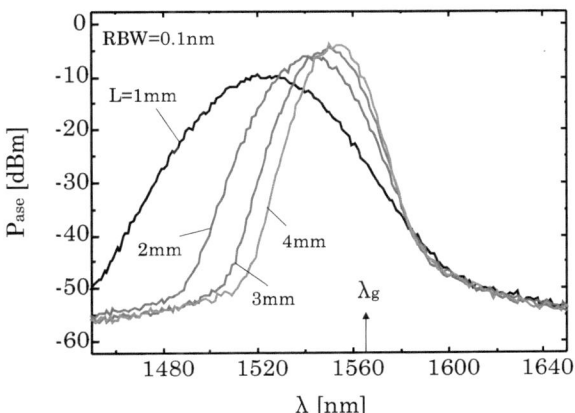

Figure G.1: Simulated ASE spectren for SOAs with different lengths (I_B = 190 mA/mm). The parameters are given in Table G.1

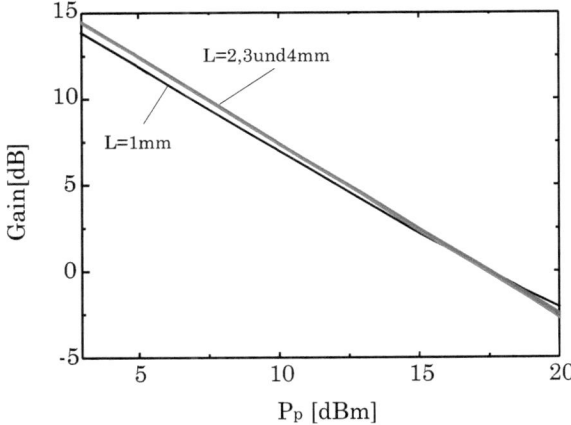

Figure G.2: Simulated gain for SOAs with different lengths, all simulated at the ASE peak wavelength (I_B = 190 mA/mm). The parameters are given in Table G.1

Appendix H

Analytical Solutions for FWM in HNLF

In this appendix, the approximate analytical solutions for the parametrically amplified and wavelength converted waves in the HNLF are calculated in a similar way as in [85] and [50, p. 372].

H.1 Single Pump FWM

Starting point is the NLS equation 2.39. Dispersive effects and the fiber attenuation shall be neglected, such that it takes the form

$$\frac{\partial}{\partial z} A(z) = i\gamma |A(z)|^2 A(z). \tag{H.1}$$

Then, the ansatz 2.35 for three input waves is chosen. Thereby, $\Omega_1 = \Omega_s$ shall be the input signal frequency, $\Omega_2 = \Omega_i$ shall be the idler frequency and $\Omega_3 = \Omega_p$ shall be the pump frequency. Furthermore, the frequency relation $\Omega_i + \Omega_s = 2\Omega_p$ shall be fulfilled. Inserting the ansatz into the right hand side of Eq. H.1 leads to Eq. SPM/SGM. Sorting for the terms with frequencies Ω_s, Ω_i and Ω_p gives the following three coupled equations for the single-pump process,

$$\frac{dA_s}{dz} = i\gamma \left(|A_s|^2 + 2|A_i|^2 + 2|A_p|^2 \right) A_s + i\gamma A_p^2 A_i^* e^{i(2B_p - B_i - B_s)z} \tag{H.2}$$

$$\frac{dA_i}{dz} = i\gamma \left(|A_i|^2 + 2|A_s|^2 + 2|A_p|^2 \right) A_i + i\gamma A_p^2 A_s^* e^{i(2B_p - B_s - B_i)z} \tag{H.3}$$

$$\frac{dA_p}{dz} = i\gamma \left(|A_p|^2 + 2|A_s|^2 + 2|A_i|^2 \right) A_p + 2i\gamma A_s A_i A_p^* e^{i(B_s + B_i - 2B_p)z} \tag{H.4}$$

In the following, the pump wave shall be much more intense than the input signal and the idler and remains undepleted during the FWM process. With $A_p(0) = \sqrt{P_p} e^{i\phi_p}$, Eq. H.4 yields the simple solution

$$A_p(z) = \sqrt{P_p} e^{i\phi_p} e^{i\gamma P_p z} \tag{H.5}$$

169

Inserting the solution for the pump wave Eq. H.5 into Eqs. H.2 and H.3 for the input signal and the idler yields

$$\frac{dA_s}{dz} = 2i\gamma P_p A_s + i\gamma P_p A_i^* e^{2i\phi_p} e^{i(2B_p - B_i - B_s + 2\gamma P_p)z} \tag{H.6}$$

$$\frac{dA_i}{dz} = 2i\gamma P_p A_i + i\gamma P_p A_s^* e^{2i\phi_p} e^{i(2B_p - B_s - B_i + 2\gamma P_p)z} \tag{H.7}$$

Applying the transformations

$$A_s = \check{A}_s e^{2i\gamma P_p z} \tag{H.8}$$

$$A_i = \check{A}_i e^{2i\gamma P_p z} e^{2i\phi_p} \tag{H.9}$$

gives

$$\frac{d\check{A}_s}{dz} = i\gamma P_p \check{A}_i^* e^{-i\kappa_{sp} z} \tag{H.10}$$

$$\frac{d\check{A}_i}{dz} = i\gamma P_p \check{A}_s^* e^{-i\kappa_{sp} z} \tag{H.11}$$

with the phase mismatch parameter

$$\kappa_{sp} = \Delta B_{sp} + 2\gamma P_p \tag{H.12}$$

and the linear phase mismatch

$$\Delta B_{sp} = B_s + B_i - 2B_p. \tag{H.13}$$

This solution of this set of linear differential equations is given by

$$\check{A}_s(z) = \check{A}_s(0)\left[\cosh(g_{sp} z) + i\frac{\kappa_{sp}}{2g_{sp}}\sinh(g_{sp} z)\right] e^{-i\kappa_{sp} z/2} \tag{H.14}$$

$$\check{A}_i(z) = \check{A}_s(0)^* \frac{i\gamma P_p}{g_{sp}}\sinh(g_{sp} z)\, e^{-i\kappa_{sp} z/2}. \tag{H.15}$$

g_{sp} is defined by

$$g_{sp}^2 = \gamma^2 P_p^2 - \frac{\kappa_{sp}^2}{4} \tag{H.16}$$

Re-applying the transformations given by Eqs. H.8 gives the final solutions

$$A_s(z) = A_s(0)\left[\cosh(g_{sp} z) + i\frac{\kappa_{sp}}{2g_{sp}}\sinh(g_{sp} z)\right] e^{-i\Delta B_{sp} z/2} e^{i\gamma P_p z} \tag{H.17}$$

$$A_i(z) = A_s(0)^* \frac{i\gamma P_p}{g_{sp}}\sinh(g_{sp} z)\, e^{-i\Delta B_{sp} z/2}\, e^{i\gamma P_p z}\, e^{2i\phi_p}. \tag{H.18}$$

The power gain G_s and the conversion efficiency G_i are given by

$$G_s^{sp} = \frac{|A_s(z)|^2}{|A_s(0)|^2} = 1 + \frac{\gamma^2 P_p^2}{g_{sp}^2}\sinh^2(g_{sp} z) \tag{H.19}$$

$$G_i^{sp} = \frac{|A_i(z)|^2}{|A_s(0)|^2} = G_s - 1 \tag{H.20}$$

The solutions for perfect phase matching, $\kappa_{sp} = 0$, are of great practical interest because the gain is maximal in this case. The field gain and the field conversion efficiency are then given by

$$\mathcal{G}_s^{sp} = \left|\frac{A_s(z)}{A_s(0)}\right|_{\kappa_{sp}=0} = \cosh(\gamma P_p z) \tag{H.21}$$

$$\mathcal{G}_i^{sp} = \left|\frac{A_i(z)}{A_s(0)}\right|_{\kappa_{sp}=0} = \sinh(\gamma P_p z). \tag{H.22}$$

The phase shift due to the FOPA for the signal and the idler is given by

$$\vartheta_s^{sp} = \Im\left\{\ln\left(\frac{A_s(z)}{A_s(0)}\right)\right\}_{\kappa_{sp}=0} = -\Delta B_{sp} z/2 + \gamma P_p z \tag{H.23}$$

$$\vartheta_i^{sp} = \Im\left\{\ln\left(\frac{A_i(z)}{A_s(0)^*}\right)\right\}_{\kappa_{sp}=0} = \pi/2 - \Delta B_{sp} z/2 + \gamma P_p z + 2\phi_p \tag{H.24}$$

H.2 Dual Pump FWM

The starting point is also Eq. H.1. However, an ansatz with four waves must be chosen,

$$A(z,T) = \sum_{l=1}^{4} A_l(z,T)\, e^{i(B_l z - \Omega_l t)}. \tag{H.25}$$

Insertion into Eq. H.1 and sorting the terms after frequency gives four coupled differential equations,

$$\frac{dA_1}{dz} = i\gamma\left(|A_1|^2 + 2|A_2|^2 + 2|A_3|^2 + 2|A_4|^2\right)A_s + 2i\gamma A_3 A_4 A_2^* e^{i(B_3+B_4-B_1-B_2)z} \tag{H.26}$$

$$\frac{dA_2}{dz} = i\gamma\left(|A_2|^2 + 2|A_1|^2 + 2|A_3|^2 + 2|A_4|^2\right)A_s + 2i\gamma A_3 A_4 A_1^* e^{i(B_3+B_4-B_1-B_2)z} \tag{H.27}$$

$$\frac{dA_3}{dz} = i\gamma\left(|A_3|^2 + 2|A_1|^2 + 2|A_2|^2 + 2|A_4|^2\right)A_s + 2i\gamma A_1 A_2 A_4^* e^{i(B_1+B_2-B_3-B_4)z} \tag{H.28}$$

$$\frac{dA_4}{dz} = i\gamma\left(|A_4|^2 + 2|A_1|^2 + 2|A_2|^2 + 2|A_3|^2\right)A_s + 2i\gamma A_1 A_2 A_3^* e^{i(B_1+B_2-B_3-B_4)z}. \tag{H.29}$$

H.2.1 Phase Conjugation

Here, $\Omega_1 = \Omega_{p1}$, $\Omega_2 = \Omega_{p2}$, $\Omega_3 = \Omega_s$ and $\Omega_4 = \Omega_i$ as well as $\Omega_{p1} + \Omega_{p2} = \Omega_s + \Omega_i$ hold. The pump waves shall be much more intense than the input signal and the idler wave and shall not be depleted during the FWM process. Then, Eqs. H.26 and H.27 yield the solutions for the pump waves,

$$A_{p1}(z) = \sqrt{P_{p1}}\, e^{i\phi_{p1}} e^{i\gamma(P_{p1}+2P_{p2})z} \tag{H.30}$$

$$A_{p2}(z) = \sqrt{P_{p2}}\, e^{i\phi_{p2}} e^{i\gamma(2P_{p1}+P_{p2})z}. \tag{H.31}$$

Insertion of the solutions for the pumps into the equations for the signal and the idler, Eqs. H.28 and H.29, yields

$$\frac{dA_s}{dz} = 2i\gamma(P_{p1}+P_{p2})A_s + 2i\gamma\sqrt{P_{p1}P_{p2}}A_i^* e^{i(\phi_{p1}+\phi_{p2})} e^{i(-\Delta B_{pc}+3\gamma(P_{p1}+P_{p2}))z} \quad (H.32)$$

$$\frac{dA_i}{dz} = 2i\gamma(P_{p1}+P_{p2})A_i + 2i\gamma\sqrt{P_{p1}P_{p2}}A_s^* e^{i(\phi_{p1}+\phi_{p2})} e^{i(-\Delta B_{pc}+3\gamma(P_{p1}+P_{p2}))z} \quad (H.33)$$

with the linear phase mismatch $\Delta B_{pc} = B_s + B_i - B_{p1} - B_{p2}$. Applying the transformation

$$A_s = \check{A}_s e^{2i\gamma(P_{p1}+P_{p2})z} \quad (H.34)$$

$$A_i = \check{A}_i e^{2i\gamma(P_{p1}+P_{p2})z} e^{i(\phi_{p1}+\phi_{p2})} \quad (H.35)$$

yields

$$\frac{d\check{A}_s}{dz} = 2i\gamma\sqrt{P_{p1}P_{p2}}\check{A}_i^* e^{-i\kappa_{pc}z} \quad (H.36)$$

$$\frac{d\check{A}_i}{dz} = 2i\gamma\sqrt{P_{p1}P_{p2}}\check{A}_s^* e^{-i\kappa_{pc}z} \quad (H.37)$$

with the phase mismatch parameter

$$\kappa_{pc} = \Delta B_{pc} + \gamma(P_{p1}+P_{p2}). \quad (H.38)$$

The solutions of the two coupled linear differential equations are given by

$$\check{A}_s(z) = \check{A}_s(0)\left[\cosh(g_{pc}z) + i\frac{\kappa_{pc}}{2g_{pc}}\sinh(g_{pc}z)\right] e^{-i\kappa_{pc}z/2} \quad (H.39)$$

$$\check{A}_i(z) = \check{A}_s(0)^* \frac{2i\gamma\sqrt{P_{p1}P_{p2}}}{g_{pc}}\sinh(g_{pc}z)\, e^{-i\kappa_{pc}z/2}. \quad (H.40)$$

g_{pc} is defined by

$$g_{pc}^2 = 4\gamma^2 P_{p1}P_{p2} - \frac{\kappa_{pc}^2}{4} \quad (H.41)$$

Re-applying the transformations given by Eqs. H.34 gives the final solutions

$$A_s(z) = A_s(0)\left[\cosh(g_{pc}z) + i\frac{\kappa_{pc}}{2g_{pc}}\sinh(g_{pc}z)\right] e^{-i\Delta B_{pc}z/2} e^{i3/2\gamma(P_{p1}+P_{p2})z} \quad (H.42)$$

$$A_i(z) = A_s(0)^* \frac{2i\gamma\sqrt{P_{p1}P_{p2}}}{g_{pc}}\sinh(g_{pc}z)\, e^{-i\Delta B_{pc}z/2}\, e^{i3/2\gamma(P_{p1}+P_{p2})z}\, e^{i(\phi_{p1}+\phi_{p2})}. \quad (H.43)$$

The power gain G_s and the conversion efficiency G_i are given by

$$G_s^{pc} = \frac{|A_s(z)|^2}{|A_s(0)|^2} = 1 + \frac{4\gamma^2 P_{p1}P_{p2}}{g_{pc}^2}\sinh^2(g_{pc}z) \quad (H.44)$$

$$G_i^{pc} = \frac{|A_i(z)|^2}{|A_s(0)|^2} = G_s - 1 \quad (H.45)$$

For the case of perfect phase matching, $\kappa_{pc} = 0$, the field gain and the field conversion efficiency are given by

$$\mathcal{G}_s^{pc} = \left|\frac{A_s(z)}{A_s(0)}\right|_{\kappa_{pc}=0} = \cosh(2\gamma\sqrt{P_{p1}P_{p2}}z) \tag{H.46}$$

$$\mathcal{G}_i^{pc} = \left|\frac{A_i(z)}{A_s(0)}\right|_{\kappa_{pc}=0} = \sinh(2\gamma\sqrt{P_{p1}P_{p2}}z). \tag{H.47}$$

The phase shift due to the FOPA for the signal and the idler is then given by

$$\vartheta_s^{pc} = \Im\left\{\ln\left(\frac{A_s(z)}{A_s(0)}\right)\right\}_{\kappa_{pc}=0} = -\Delta B_{pc}z/2 + \frac{3}{2}\gamma(P_{p1}+P_{p2})z \tag{H.48}$$

$$\vartheta_i^{pc} = \Im\left\{\ln\left(\frac{A_i(z)}{A_s(0)^*}\right)\right\}_{\kappa_{pc}=0} = \pi/2 - \Delta B_{pc}z/2 + \frac{3}{2}\gamma(P_{p1}+P_{p2})z + \phi_{p1} + \phi_{p2} \tag{H.49}$$

H.2.2 Frequency conversion

Here, $\Omega_1 = \Omega_{p1}$, $\Omega_2 = \Omega_s$, $\Omega_3 = \Omega_{p2}$ and $\Omega_4 = \Omega_i$ as well as $\Omega_{p2} + \Omega_s = \Omega_{p1} + \Omega_i$ hold. The pump waves shall be much more intense than the input signal and the idler wave and shall not be depleted during the FWM process. Then, the solutions for the pump waves are given by Eqs. H.30. Insertion in the equations for the signal and the idler, Eqs. H.27 and H.29 yields

$$\frac{dA_s}{dz} = 2i\gamma(P_{p1}+P_{p2})A_s + 2i\gamma\sqrt{P_{p1}P_{p2}}A_i e^{i(\phi_{p1}-\phi_{p2})}e^{i(\Delta B_{fc}+\gamma(P_{p2}-P_{p1}))z} \tag{H.50}$$

$$\frac{dA_i}{dz} = 2i\gamma(P_{p1}+P_{p2})A_i + 2i\gamma\sqrt{P_{p1}P_{p2}}A_s e^{i(\phi_{p2}-\phi_{p1})}e^{i(-\Delta B_{fc}+\gamma(P_{p1}-P_{p2}))z} \tag{H.51}$$

with the linear phase mismatch $\Delta B_{fc} = B_{p1} + B_i - B_{p2} - B_s$. Applying the transformation

$$A_s = \check{A}_s e^{2i\gamma(P_{p1}+P_{p2})z} \tag{H.52}$$

$$A_i = \check{A}_i e^{2i\gamma(P_{p1}+P_{p2})z} e^{i(\phi_{p2}-\phi_{p1})} \tag{H.53}$$

yields

$$\frac{d\check{A}_s}{dz} = 2i\gamma\sqrt{P_{p1}P_{p2}}\check{A}_i e^{i\kappa_{fc}z} \tag{H.54}$$

$$\frac{d\check{A}_i}{dz} = 2i\gamma\sqrt{P_{p1}P_{p2}}\check{A}_s e^{-i\kappa_{fc}z} \tag{H.55}$$

with the phase mismatch parameter

$$\kappa_{fc} = \Delta B_{fc} + \gamma(P_{p2}-P_{p1}). \tag{H.56}$$

The solutions of the two coupled linear differential equations are given by

$$\check{A}_s(z) = \check{A}_s(0)\left[\cos(g_{fc}z) - i\frac{\kappa_{fc}}{2g_{fc}}\sin(g_{fc}z)\right]e^{i\kappa_{fc}z/2} \tag{H.57}$$

$$\check{A}_i(z) = \check{A}_s(0)\frac{2\gamma\sqrt{P_{p1}P_{p2}}}{g_{fc}}\sin(g_{fc}z)\,e^{-i\kappa_{fc}z/2}. \tag{H.58}$$

g_{fc} is defined by

$$g_{fc}^2 = 4\gamma^2 P_{p1}P_{p2} + \frac{\kappa_{fc}^2}{4}. \tag{H.59}$$

Re-applying the transformations given by Eqs. H.52 gives the final solutions

$$A_s(z) = A_s(0)\left[\cos(g_{fc}z) - i\frac{\kappa_{fc}}{2g_{fc}}\sin(g_{fc}z)\right]e^{i\Delta B_{fc}z/2}e^{i\gamma(3/2P_{p1}+5/2P_{p2})z} \tag{H.60}$$

$$A_i(z) = A_s(0)\frac{2\gamma\sqrt{P_{p1}P_{p2}}}{g_{fc}}\sin(g_{fc}z)\,e^{-i\Delta B_{fc}z/2}\,e^{i\gamma(5/2P_{p1}+3/2P_{p2})z}\,e^{i(\phi_{p1}-\phi_{p2})}. \tag{H.61}$$

The power gain G_s and the conversion efficiency G_i are given by

$$G_s^{fc} = \frac{|A_s(z)|^2}{|A_s(0)|^2} = 1 - \frac{4\gamma^2 P_{p1}P_{p2}}{g_{fc}^2}\sin^2(g_{fc}z) \tag{H.62}$$

$$G_i^{fc} = \frac{|A_i(z)|^2}{|A_s(0)|^2} = 1 - G_s \tag{H.63}$$

For the case of perfect phase matching, $\kappa_{fc} = 0$, the field gain and the field conversion efficiency are given by

$$\mathscr{G}_s^{fc} = \left|\frac{A_s(z)}{A_s(0)}\right|_{\kappa_{fc}=0} = \cos(2\gamma\sqrt{P_{p1}P_{p2}}z) \tag{H.64}$$

$$\mathscr{G}_i^{fc} = \left|\frac{A_i(z)}{A_s(0)}\right|_{\kappa_{fc}=0} = \sin(2\gamma\sqrt{P_{p1}P_{p2}}z). \tag{H.65}$$

The phase shift due to the FOPA for the signal and the idler is then given by

$$\vartheta_s^{fc} = \Im\left\{\ln\left(\frac{A_s(z)}{A_s(0)}\right)\right\}_{\kappa_{fc}=0} = \Delta B_{fc}z/2 + \gamma(3/2P_{p1}+5/2P_{p2})z \tag{H.66}$$

$$\vartheta_i^{fc} = \Im\left\{\ln\left(\frac{A_i(z)}{A_s(0)}\right)\right\}_{\kappa_{fc}=0} = -\Delta B_{fc}z/2 + \gamma(5/2P_{p1}+3/2P_{p2})z + \phi_{p1} - \phi_{p2}. \tag{H.67}$$

Appendix I

BER Calculation for 16-QAM

In this appendix, it is shown how the BER was calculated for 16-QAM signals in the sections 4.4.3 and 4.4.3 of this thesis. The amplitude of the FOPA output signal is given by

$$|A_s(L,t)| = \langle \mathcal{G}_s^{sp} \rangle \widetilde{\mathcal{G}}_s^{sp}(t) |A_s(0,t) + n_c(0,t)|. \tag{I.1}$$

where $A_s(0,t)$ is the complex-valued, noiseless input signal carrying the 16-QAM modulation and $n_c(0,t)$ is a complex-valued noise contribution representing the AWG noise at the FOPA input. Using $|A_s(0,t)| \gg |n_c(0,t)|$, one can write

$$\begin{aligned}|A_s(0,t) + n_c(0,t)| &= \sqrt{(A_s(0,t) + n_c(0,t))(A_s^*(0,t) + n_c^*(0,t))} \\ &\approx |A_s(0,t)| \sqrt{1 + 2\frac{\Re\{A_s^*(0,t)n_c(0,t)\}}{|A_s(0,t)|^2}} \\ &\approx |A_s(0,t)| + \underbrace{\frac{\Re\{A_s^*(0,t)n_c(0,t)\}}{|A_s(0,t)|}}_{n_{\parallel}(0,t)}\end{aligned} \tag{I.2}$$

where $n_{\parallel}(0,t)$ is the real-valued AWG noise contribution parallel (inphase) to the input signal. Insertion in Eq. I.1 leads to

$$|A_s(L,t)| = \langle \mathcal{G}_s^{sp} \rangle \widetilde{\mathcal{G}}_s^{sp}(t) \big(|A_s(0,t)| + n_{\parallel}(L,t)\big). \tag{I.3}$$

Because $\langle \widetilde{\mathcal{G}}_s^{sp}(t) \rangle \cong 1$ after Eq. 4.135, one can write

$$\widetilde{\mathcal{G}}_s^{sp}(t) = 1 + \underbrace{(\widetilde{\mathcal{G}}_s^{sp}(t) - 1)}_{\ll 1}. \tag{I.4}$$

Using this gives,

$$|A_s(L,t)| \cong \underbrace{\langle \mathcal{G}_s^{sp} \rangle |A_s(0,t)|}_{<|A_s(L,t)|>} + \underbrace{\langle \mathcal{G}_s^{sp} \rangle n_{\parallel}(0,t)}_{n_{\parallel}(L,t)} + \underbrace{\langle \mathcal{G}_s^{sp} \rangle (\widetilde{\mathcal{G}}_s^{sp}(t) - 1)|A_s(0,t)|}_{n_{nan}(t)}. \tag{I.5}$$

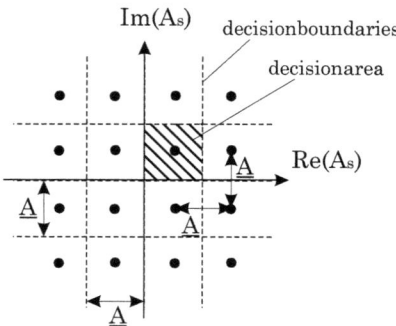

Figure I.1: Constellation of the 16-QAM format with decision boundaries used for the BER calculation.

The output amplitude $|A_s(L,t)|$ consists the noiseless amplitude $<|A_s(L,t)|>$ and two additive noise terms. The first, $n_{\|}(L,t)$ is Gaussian-distributed and shall also include AWG noise added by the FOPA due to its non-zero noise figure. The second, $n_{nan}(t)$, carries the nonlinear amplitude noise. A fourth contribution, proportional to the product $(\widetilde{\mathcal{G}}_s^{sp}(t)-1)n_{\|}(0,t)$ was neglected because it is a product of two small terms. The SNR of the output signal (taking into account only Gaussian distributed noise) is given by

$$\text{SNR}_s(L) = \frac{<P_s>}{2<n_{\|}^2(L,t)>} \tag{I.6}$$

where the average power of the 16-QAM signal is given by a summation over each of the 16 symbols,

$$<P_s> = <|A_s(L,t)|^2> = \frac{1}{16}\sum_{k=1}^{16} <|A_{16\text{QAM},k}(L)|^2>, \tag{I.7}$$

where it was assumed that each of the 16 possible symbols $A_{16\text{QAM},k}$ has the same probability to appear within a any time period. The constellation of the 16-QAM format is shown in Fig. I.1. The mean amplitudes and the powers of the symbols in the first quadrant are given in Table I.1. The symbols of the other quadrants are easily derived due to the rotational symmetry of the 16-QAM constellation. Using Table I.1, $<P_s> = 10\underline{A}^2$. To evaluate the BER, the probability distribution functions of each symbol have to be known. Using Eq. 3.14, the PDF of the Gaussian distributed part of the signal is given by

$$\text{PDF}_{<|A_{16\text{QAM},k}(L)|> + n_{\|}(L,t)}(x) = \sqrt{\frac{1}{2\pi<n_{\|}^2(L,t)>}} \exp\left(-\frac{(x - <|A_{16\text{QAM},k}(L)|>)^2}{2<n_{\|}^2(L,t)>}\right). \tag{I.8}$$

The PDF of the nonlinear amplitude noise can be calculated as follows. If a random variable X is transformed according to $Y = aX + b$, then the PDF of X

is transformed according to
$$\text{PDF}_Y(y) = \frac{1}{a}\text{PDF}_X\left(\frac{y-b}{a}\right). \tag{I.9}$$
Thus, we can use Eqs. (4.132) and (I.9) to derive the PDF of the lognormally distributed term $n_{nan}(t)$ in Eq. (I.5) which is given by
$$\text{PDF}_{n_{nan,l}}(x) = \frac{1}{<|A_{16\text{QAM},k}(L)|>}\text{PDF}_{\tilde{g}_s^{sp}}\left(\frac{x}{<|A_{16\text{QAM},k}(L)|>}+1\right).$$
The overall PDF of the l-th symbol amplitude, $<|A_{16\text{QAM},k}(L)|> + n_\|(L,t) + n_{nan}(t)$, is given by the convolution of the PDFs of the Gaussian distributed part of the signal and the PDF of the nonlinear amplitude noise,
$$\text{PDF}_{<|A_{16\text{QAM},k}(L)|>+n_\|(L,t)+n_{nan,l}(t)} = \text{PDF}_{<|A_{16\text{QAM},k}(L)|>+n_\|(L,t)} * \text{PDF}_{n_{nan,l}(t)}. \tag{I.10}$$
The PDF of the output signal phase degraded by AWG noise and nonlinear phase noise, $\text{PDF}_{\theta_{s,l}}$, is given by Eq. 3.31 where the coefficients given in Eq. 3.21 are adapted to
$$c_k = \frac{\sqrt{\pi\text{SNR}_{s,l}}}{2}e^{\text{SNR}_{s,l}/2}\left[I_{\frac{k-1}{2}}\left(\frac{\text{SNR}_{s,l}}{2}\right) + I_{\frac{k+1}{2}}\left(\frac{\text{SNR}_{s,l}}{2}\right)\right]. \tag{I.11}$$
using the symbol SNR
$$\text{SNR}_{s,l} = \text{SNR}_s \frac{<|A_{16\text{QAM},k}(L)|^2>}{<P_s>} \tag{I.12}$$
which depends on the symbol. Assuming that amplitude and phase of a signal are independent variables, the joint PDF of amplitude and the phase is given by multiplication of the individual PDFs. Then, the BER can be calculated by using Eq. 3.18,
$$\text{BER} = \frac{1}{\log_2(16)\times 16}$$
$$\times \sum_{l=1}^{16}\iint_{F_{dec,l}}\left(1-\text{PDF}_{<|A_{16\text{QAM},k}(L)|>+n_\|(L,t)+n_{nan,l}(t)}\text{PDF}_{\theta_{s,l}}\right)d|A_s|d\theta_s \tag{I.13}$$
where $F_{dec,l}$ is the area that is enclosed by the decision boundaries of the l-th symbol. Due to the rotational symmetry of the 16-QAM constellation, it is sufficient to perform the integration in the first quadrant. The used decision areas of the symbols in the first quadrant are given in Table I.1 for the ($\Re\{A_s\},\Im\{A_s\}$)-plane and shown exemplary in Fig. I.1. The term $\pm\underline{\delta}$ refers to the fact that decision boundaries have been optimized for minimal BER because of the asymmetry of the nonlinear amplitude noise PDF, $\text{PDF}_{n_{nan,l}(t)}$. For pure Gaussian noise, $\underline{\delta} = 0$ is optimal. To evaluate the integral in Eq. I.13, the decision areas have to be mapped to the ($|A_s|,\theta_s$)-plane using the relations
$$\Re\{A_s\} = |A_s|\cos(\theta_s)$$
$$\Im\{A_s\} = |A_s|\sin(\theta_s).$$

Table I.1: Symbols of the 16-QAM constellation in the first quadrant after Fig. I.1: Mean amplitudes, symbol powers and quadratic decision areas (left lower corner → right upper corner)

Symbol	Mean amplitude $A_{16\text{QAM},k}$	Symbol power $\|A_{16\text{QAM},k}\|^2$	Decision area $F_{dec,l}$
Type 1	$\frac{1}{2}\underline{A} + i\frac{1}{2}\underline{A}$	$\frac{1}{2}\underline{A}^2$	$(0,0) \to (\underline{A}\pm\underline{\delta},\underline{A}\pm\underline{\delta})$
Type 2	$\frac{3}{2}\underline{A} + i\frac{1}{2}\underline{A}$	$\frac{5}{2}\underline{A}^2$	$(\underline{A},0) \to (\infty,\underline{A}\pm\underline{\delta})$
	$\frac{1}{2}\underline{A} + i\frac{3}{2}\underline{A}$	$\frac{5}{2}\underline{A}^2$	$(0,\underline{A}) \to (\underline{A}\pm\underline{\delta},\infty)$
Type 1	$\frac{3}{2}\underline{A} + i\frac{3}{2}\underline{A}$	$\frac{9}{2}\underline{A}^2$	$(\underline{A}\pm\underline{\delta},\underline{A}\pm\underline{\delta}) \to (\infty,\infty)$

Appendix J

Phase Distortion after Carrier Phase Estimation

In this appendix, 4.96 is derived. Starting point is Eq. 4.94 which gives the complex phasor of the wavelength converted signal after the m-th power operation. Thereby, it is assumed that $m \, m_{M,\text{eff}} \ll 1$. In the next step, a running average over N samples is performed to cancel out Gaussian noise (that is not included in this calculation),

$$
\begin{aligned}
<s(k)^m>\big|_{N_{av}} &= \frac{1}{N_{av}} \sum_{l=k-(N_{av}-1)/2}^{k+(N_{av}-1)/2} s(l)^m \\
&= \frac{1}{N_{av}} \sum_{l=k-(N_{av}-1)/2}^{k+(N_{av}-1)/2} \exp\left(im \, m_{M,i} \cos(2\pi l f_M/R_s)\right) \\
&\approx \frac{1}{N_{av}} \sum_{l=k-(N_{av}-1)/2}^{k+(N_{av}-1)/2} \left[1 + im \, m_{M,i} \cos(2\pi l f_M/R_s)\right] \\
&= 1 + \frac{1}{N_{av}} \sum_{l=k-(N_{av}-1)/2}^{k+(N_{av}-1)/2} im \, m_{M,i} \cos(2\pi l f_M/R_s) \\
&\approx \exp\left(\frac{1}{N_{av}} \sum_{l=k-(N_{av}-1)/2}^{k+(N_{av}-1)/2} im \, m_{M,i} \cos(2\pi l f_M/R_s)\right).
\end{aligned} \quad (\text{J.1})
$$

The calculation shows that, for $m \, m_{M,i} \ll 1$, the averaging over the complex phasor can be replaced by an average over the phase of the phasor which can be extracted by an **unwrapping** arg-operation and a division by m,

$$
\begin{aligned}
\psi(k) &= 1/m \, \arg\left[<s(k)^m>\big|_{N_{av}}\right] \\
&= <\phi_{ppm}(k)>\big|_{N_{av}} \\
&= \frac{1}{N_{av}} \sum_{l=k-(N_{av}-1)/2}^{k+(N_{av}-1)/2} m_{M,i} \cos(2\pi l f_M/R_s)
\end{aligned} \quad (\text{J.2})
$$

Using the window function $\Pi_{N_{av}}(k)$ defined as

$$\Pi_{N_{av}}(k) = \begin{cases} 1 & |k| \le (N_{av}-1)/2 \\ 0 & \text{otherwise} \end{cases} \tag{J.3}$$

the running average can be interpreted as a discrete convolution,

$$\psi(k) = \frac{1}{N_{av}} \sum_{l=-\infty}^{+\infty} \phi_{ppm}(l) \, \Pi_{N_{av}}(k-l), \tag{J.4}$$

which can be evaluated by a discrete Fourier transform:

$$\mathcal{F}\{\psi(k)\} = \frac{1}{N_{av}} \, \mathcal{F}\{\phi_{ppm}(k)\} \cdot \mathcal{F}\{\Pi_{N_{av}}(k)\} \tag{J.5}$$

The discrete Fourier transform of $\phi_{ppm}(k)$ is given by two delta function combs,

$$\mathcal{F}\{\phi_{ppm}(k)\} = m_{M,i} \, \pi \sum_{l=-\infty}^{+\infty} \delta(\Omega - \Omega_M - 2\pi l) + \delta(\Omega + \Omega_M - 2\pi l) \tag{J.6}$$

with $\Omega = 2\pi f/R_s$ and $\Omega_M = 2\pi f_M/R_s$ while the discrete Fourier transform of the window function is given by

$$\mathcal{F}\{\Pi_{N_{av}}(k)\} = \frac{\sin(N_{av} \, \Omega/2)}{\sin(\Omega/2)} \tag{J.7}$$

The product gives

$$\mathcal{F}\{\psi(k)\} = \frac{m_{M,i} \, \pi}{N_{av}} \sum_{l=-\infty}^{+\infty} \frac{\sin(N_{av} \, (\Omega_M + 2\pi l)/2)}{\sin((\Omega_M + 2\pi l)/2)} \delta(\Omega - \Omega_M - 2\pi l)$$
$$+ \frac{\sin(N_{av} \, (-\Omega_M + 2\pi l)/2)}{\sin((-\Omega_M + 2\pi l)/2)} \delta(\Omega + \Omega_M - 2\pi l) \tag{J.8}$$

With $\sin(x+y) = \sin(x)\cos(y) + \cos(x)\sin(y)$ follows

$$\mathcal{F}\{\psi(k)\} =$$
$$\frac{m_{M,i} \, \pi}{N_{av}} \sum_{l=-\infty}^{+\infty} \frac{\sin(N_{av} \, \Omega_M/2)\cos(\pi l N_{av}) + \cos(N_{av} \, \Omega_M/2)\sin(\pi l N_{av})}{\sin(\Omega_M/2)\cos(\pi l) + \cos(\Omega_M/2)\sin(\pi l)}$$
$$\times \delta(\Omega - \Omega_M - 2\pi l)$$
$$+ \frac{\sin(-N_{av} \, \Omega_M/2)\cos(\pi l N_{av}) + \cos(N_{av} \, \Omega_M/2)\sin(\pi l N_{av})}{\sin(-\Omega_M/2)\cos(\pi l) + \cos(\Omega_M/2)\sin(\pi l)}$$
$$\times \delta(\Omega + \Omega_M - 2\pi l)$$

We know that l is an integer and N_{av} is an odd integer:

$$\sin(\pi l N_{av}) = 0$$
$$\sin(\pi l) = 0$$
$$\cos(\pi l N_{av}) = \begin{cases} +1 & l \text{ even} \\ -1 & l \text{ odd} \end{cases}$$
$$\cos(\pi l) = \begin{cases} +1 & l \text{ even} \\ -1 & l \text{ odd} \end{cases}$$

Simplifying gives

$$\mathcal{F}\{\psi(k)\} = m_{M,i} \frac{\sin(N_{av}\,\Omega_M/2)}{N_{av}\,\sin(\Omega_M/2)} \pi \sum_{l=-\infty}^{+\infty} \delta(\Omega - \Omega_M - 2\pi l) + \delta(\Omega + \Omega_M - 2\pi l)$$

Now, we apply the inverse Fourier transform and end up with

$$\psi(k) = m_{M,i} \frac{\sin(N_{av}\,\Omega_M/2)}{N_{av}\,\sin(\Omega_M/2)} \cos(k\Omega_M)$$

$$= m_{M,i} \frac{\sin\left(\frac{\pi N_{av} f_M}{R_s}\right)}{N_{av}\,\sin\left(\frac{\pi f_M}{R_s}\right)} \cos\left(k\frac{2\pi f_M}{R_s}\right)$$

$$= \frac{\sin\left(\frac{\pi N_{av} f_M}{R_s}\right)}{N_{av}\,\sin\left(\frac{\pi f_M}{R_s}\right)} \phi_{ppm}(k). \tag{J.9}$$

Appendix K

Quadratic interpolation

In this appendix, Eqs. 4.99 and 4.100 are derived. The derivation follows the quadratically interpolated FFT method (QIFFT) [116] that is used to estimate parameters of sinusoidals in audio technology. The general form of the quadratic equation is given by

$$f(x) = -a_0(x - x_0)^2 + y_0 \qquad (K.1)$$

The value pair of the maximum, (x_0, y_0), is unknown and shall be determined. The three interpolation value pairs are given by $(-\Delta x, \underline{a})$, $(0, \underline{b})$ and $(\Delta x, \underline{c})$. Insertion into the general form gives three equations,

$$\begin{aligned}
\underline{a} &= -a_0 \Delta x^2 - 2a_0 \Delta x x_0 - a_0 x_0^2 + y_0 \\
\underline{b} &= -a_0 x_0^2 + y_0 \\
\underline{c} &= -a_0 \Delta x^2 + 2a_0 \Delta x x_0 - a_0 x_0^2 + y_0.
\end{aligned} \qquad (K.2)$$

Then, the maximum is given by

$$x_0 = \frac{\Delta x}{2} \frac{\underline{a} - \underline{c}}{\underline{a} - 2\underline{b} + \underline{c}} \qquad (K.3)$$

$$y_0 = \underline{b} + \frac{x_0}{4\Delta x}(\underline{a} - \underline{c}). \qquad (K.4)$$

Appendix L

List of Acronyms

Mathematical symbols

∇	Nabla operator
$\frac{\partial}{\partial n}$	Normal derivative
A	Slowly varying complex envelope of the scalar electrical field
$\tilde{A}$	Fourier transform of the slowly varying complex electrical field envelope
$\underline{A}$	Parameter for 16-QAM BER calculation
$A^{\pm}$	Slowly varying complex field envelope traveling in $\pm z$ direction
A_{eff}	Effective mode area
A_i	Converted signal (idler) complex amplitude
A_{nr}	Non-radiative recombination coefficient
A_p	Pump complex amplitude
A_{rec}	Signal complex amplitude
A_s	Signal complex amplitude

A_{SE}	Slowly varying complex spontaneous emission field
A_0	Initial signal complex amplitude
a	Pulse shape
$\bar{a}$	Linear gain fitting coefficient
$\hat{a}$	Parameter of the FIR filter fitting procedure
$\tilde{a}$	Distorted pulse shape
$\underline{a}$	Parameter of the quadratic interpolation
a_0	Parameter of the quadratic interpolation
a_N	SOA differential gain
B	Difference wavenumber
$\underline{B}$	Normalized wavenumber
$\vec{B}$	Vectorial magnetic flux density
B_N	Optical noise bandwidth
B_{ref}	Reference noise bandwidth
B_{sp}	Spontaneous radiative recombination coefficient
BER	Bit error ratio
$\bar{b}$	Gain peak shift fitting coefficient
$\hat{b}$	Parameter of the FIR filter fitting procedure
$\underline{b}$	Parameter of the quadratic interpolation
b_0	Peak frequency shift coefficient
C	Normalization constant for the electrical field
C_{Auger}	Auger recombination coefficient

$\underline{c}$	Parameter of the quadratic interpolation
c_l	l-th coefficient of the BER function of m-(D)PSK signals
c_0	Vacuum speed of light
$c_{1,l}, c_{2,l}$	FIR filter coefficients for the l-th segment
D	HNLF cladding diameter
$\hat{D}$	Linear NLSE operator
$\underline{D}$	HNLF dispersion
$\vec{D}$	Vectorial electric flux density
D_p	PMD coefficient
d	Core diameter
$\hat{d}$	Parameter of the FIR filter fitting procedure
d_w	waveguide thickness
E	Scalar electrical field
$\hat{E}$	Complex envelope of the scalar electrical field
$\tilde{E}$	Fourier transform of the scalar electrical field
$\underline{E}$	Carrier energy
$\vec{E}$	Vectorial electrical field
E_c	Band edge energy of the conduction band
E_{fc}	Quasi fermi energy of the conduction band
E_{fv}	Quasi fermi energy of the valence band
E_{gap}	Bandgap energy
$|E_s|^2$	Saturation field intensity

E_v	Band edge energy of the valence band
F	Transverse electrical field profile
F_k	Integration area for BER calculations
F_0	Transverse electrical field profile (0th order)
f	Frequency
f_b	Parameter of the quadratic interpolation
f_c	Fermi distribution in the conduction band
f_g	SOA band gap frequency
f_{sin}	Modulation frequency for pump amplitude modulation
f_p	Pump wave frequency
f_{zd}	Zero-dispersion frequency
$f_1, ..., f_M$	Pump phase modulation frequencies
$\tilde{f}_1, ..., \tilde{f}_M$	Rough estimates of the pump phase modulation frequencies
G_{FIR}	Complex transfer function of the FIR filter
G_i	Power conversion efficiency
$\mathcal{G}_i$	Field conversion efficiency
$\tilde{\mathcal{G}}_i$	Field conversion efficiency fluctuation
G_l	SOA complex gain in the l-th segment
G_s	Signal power gain
$\mathcal{G}_s$	Field gain

g	SOA gain coefficient
$g^{\pm}$	Effective SOA gain coefficient in $\pm z$ direction
g_B	Brillouin gain coefficient
g_{CDP}	SOA gain coefficient due to carrier density pulsations
g_{CH}	SOA gain coefficient due to carrier heating
g_{fc}	Gain coefficient of the frequency conversion AOWC process
g_{FCA}	SOA gain coefficient due to free carrier absorption
g_p	SOA peak gain coefficient
$g_{p,2}$	SOA peak gain quadratic coefficient
$g_{p,3}$	SOA peak gain cubic coefficient
g_{pc}	Gain coefficient of the phase conjugation AOWC process
g_R	Raman peak gain coefficient
g_{sp}	Gain coefficient of the single pump AOWC process
g_{SHB}	SOA gain coefficient due to spectral hole burning
g_{TPA}	SOA gain coefficient due to two photon absorption
$\vec{H}$	Vectorial magnetic field
H_{npn}	SOA nonlinear phase noise transfer function
h	Planck constant
$\hbar$	Planck constant times 2π
h_s	Step size

I_B	SOA bias current
I_{rec}	Received current
$I_{I,k}$	Signal in-phase component current of the k-th symbol
$I_{\text{rec},I}$	Signal in-phase component current
$I_{\text{rec},Q}$	Signal quadrature component current
$I_{Q,k}$	Signal quadrature component current of the k-th symbol
$\vec{J}$	Vectorial current density
J_n	Bessel function of order n
K_n	Modified Hankel function of order n
k_B	Boltzmann constant
k_e	Electron wavenumber
k_p	Proportionality factor between power and photon density
k_0	Vacuum propagation constant
L	Fiber/SOA length
L_s	Transmission span loss
l	Azimuthal order of fiber mode
M	Number of sinusoidal phase modulation tones
m	PSK order
m_p	Pump amplitude modulation index
m_s	Signal amplitude modulation index
$m_1, ..., m_M$	Pump phase modulation index
$m_{1,eff}$	Effective phase modulation index

$m_{1,i}, ..., m_{M,i}$	Idler phase modulation index
m'	Modulation index multiplier for cascaded wavelength conversions
m'_{BC}	Best case modulation index multiplier for cascaded wavelength conversions
m'_{WC}	Worst case modulation index multiplier for cascaded wavelength conversions
NF	Noise figure
N	Total carrier density
N_{av}	Number of averaged symbols
$\hat{N}$	Nonlinear NLSE operator
N_c	Number of cascaded wavelength conversions
N_s	Number of transmission spans
N_{tr}	Transparency carrier density
N_{un}	Unsaturated carrier density
$\underline{n}$	Intensity dependent refractive index
n_c	Complex additive white Gaussian noise
n_{cl}	Cladding refractive index
n_{co}	Core refractive index
n_I	AWG noise in the in-phase component
n_{nan}	Nonlinear amplitude noise
n_Q	AWG noise in the quadrature component
n_{sp}	Inversion factor
n_0	Linear refractive index

n_2	Third order nonlinear refractive index
OSNR	Optical signal-to-noise ratio
OSNR_a	Available optical signal-to-noise ratio
OSNR_{pen}	Optical signal-to-noise ratio penalty
OSNR_r	Required optical signal-to-noise ratio
$\tilde{P}$	Fourier transform of the scalar electrical polarization
$\vec{P}$	Vectorial electrical polarization
$P^{(1)}$	First order (linear) electrical polarization
$P^{(3)}$	Third order (nonlinear) electrical polarization
P_{ASE}	ASE power in optical reference bandwidth
P_{av}	Average signal power
P_c	Pump power spectrum
P_{in}	Launch power (at the transmitter)
P_k	Power of the k-th symbol
$\tilde{P}_k$	Distorted power of the k-th symbol
P_{LO}	Local oscillator power
P_p	Pump power
P_r	Steady state resonant electrical polarization
P_{sat}	SOA saturation power
P_{st}	Stokes power
$P_{th,SBS}$	Threshold power for stimulated Brillouin scattering

$P_{th,SRS}$	Threshold power for stimulated Raman scattering
PDF_{A_k}	PDF of the k-th electrical field symbol
$\text{PDF}_{\Delta\tilde{\phi}_k}$	PDF of the k-th distorted differential symbol phase
$\text{PDF}_{\Delta\phi^{nl}}$	PDF of the differential nonlinear phase noise
$\text{PDF}_{\tilde{\phi}_k}$	PDF of the k-th distorted symbol phase
p	Radial order of fiber mode
p_{se}	Power spectral density of spontaneous Brillouin scattering
p_{st}	Stokes power spectral density
q	Electron charge
R	SOA recombination term
R_s	Symbol rate
RIN	Relative intensity noise
r	radius
$\vec{r}$	Spatial coordinate vector
$S^{\pm}$	Effective photon density traveling in ± direction
$\underline{S}_D$	Dispersion slope
SER_{A_k}	Symbol error rate of the k-th symbol
SNR_p	Pump SNR
SNR_s	Signal SNR
SPR	SOA signal-to-pump ratio
$\tilde{s}_k$	Distorted symbol normalized to its amplitude after digital equalization

T	Retarded time
T_0	Temperature
T_c	Carrier temperature
T_L	Lattice temperature
T_M	Period of maximum pump phase modulation frequency
T_s	Symbol period
t, t'	Time
t_k	Sampling instant of the k-th symbol
V	SOA active zone volume
$\underline{V}$	Normalized frequency
$\underline{V}_w$	Waveguide parameter
v_G	Group velocity
w	Effective mode radius
w_w	Active region width
X_k	Complex symbol phasor after digital equalizing
X_k	Undistorted symbol phasor after digital equalizing
X_1, X_2	In-phase and quadrature component of a complex phasor
$\underline{x}$	Mole fraction for Gallium
x_1, x_2	Gaussian distributed variables with unit variance
Y_c	Cross-correlation between Y_1 and Y_2
Y_1	Power spectrum of the idler symbols raised to the m-th power
Y_2	Y_1 mirrored at the y-axis

$\underline{y}$	Mole fraction for Arsenide
z, z'	
z_0	Zero gain frequency shift coefficient
α	Intensity dependent absorption coefficient
$\alpha_{absorption}$	Fiber attenuation coefficient due to absorption
$\alpha_{bending}$	Fiber attenuation coefficient due to bending
α_H	Henry factor of the SOA
$\alpha_{H,CDP}$	Henry factor for carrier density pulsations
$\alpha_{H,CH}$	Henry factor for carrier heating
$\alpha_{H,FCA}$	Henry factor for free carrier absorption
$\alpha_{H,SHB}$	Henry factor for spectral hole burning
$\alpha_{H,TPA}$	Henry factor for two photon absorption
α_{int}	SOA internal losses
$\alpha_{scattering}$	Fiber attenuation coefficient due to scattering
α_0	Linear absorption coefficient
α_2	Third-order nonlinear absorption coefficient
β	Propagation constant
$\tilde{\beta}$	Separation constant
β_i	Converted signal (idler) phase modulation index

β_n	Taylor expansion coefficients to the propagation constant
β_s	Signal phase modulation index
β_{TPA}	Two photon absorption coefficient
β_0	Propagation constant at the optical reference frequency
Γ	Confinement factor
Γ_{TPA}	Confinement factor for two photon absorption
γ	Real-valued nonlinear coefficient
$\tilde{\gamma}$	Complex nonlinear coefficient
Δ	Laplace operator
ΔA_p	Noise distortion of the electrical field of the pump
ΔB	Linear phase mismatch
ΔF	Perturbation of the fiber transversal mode profile
Δf	Frequency separation between signal and converted signal (idler)
Δf_s	Frequency separation between signal and zero-dispersion frequency
Δf_{CCF}	Cross-correlation frequency offset
Δf_{FFT}	FFT frequency resolution
Δn	Relative index step
ΔP_p	Pump power fluctuation
ΔP_{th}	Increase in SBS threshold power
Δ_T	Transverse Laplace operator
Δt	Time step / Sampling interval
Δz	SOA segment length

$\Delta\beta$	Perturbation of the propagation constant
$\Delta\epsilon$	Complex perturbation of the relative permittivity
$\Delta\zeta$	Perturbation of the separation constant
$\Delta\lambda$	Wavelength separation between signal and pump
$\Delta\nu_B$	Brillouin gain bandwidth
$\Delta\nu_L$	Laser linewidth
$\Delta\nu_{L,DD}$	Converted signal linewidth at the direct detection receiver
$\Delta\nu_{L,CD}$	Converted signal linewidth after coherent detection
$\Delta\Phi_{nl}$	Differential nonlinear phase distortion
$\Delta\phi_k$	Differential symbol phase
$\Delta\tilde{\phi}_k$	Distorted differential symbol phase
$\Delta\phi_{ppm}$	Differential phase distortion due to pump phase modulation
$\Delta\tau$	Delay difference for the two pump waves
$\Delta\tau_0$	Group delay difference for the two pump waves
$\Delta\omega_{LO}$	Frequency separation between received signal and local oscillator
δ	Dirac impulse function
δ_p	Perturbation parameter
$\tilde{\epsilon}$	Complex relative permittivity

ϵ_b	(Real valued) background relative permittivity
ϵ_{CH}	Gain compression factor for carrier heating
ϵ_{FCA}	Gain compression factor for free carrier absorption
ϵ_{SHB}	Gain compression factor for spectral hole burning
ϵ_{TPA}	Gain compression factor for two photon absorption
ϵ_0	Vacuum permittivity
ζ	Separation constant
ζ_β	Unknown phase shift
ζ_0	Unknown phase shift
$\theta_a, \theta_b, \theta_c$	Parameters for the quadratic interpolation
θ_e	Deterministic phase error
θ_n	Phase of the n-th pump phase modulation contribution for cascaded wavelength conversions
ϑ	Phase
ϑ_i	Idler phase shift after the FOPA / SOA
ϑ_s	Signal phase shift after the FOPA / SOA
ϑ_{un}	SOA phase shift for the unsaturated carrier density
Θ_i	Idler phase shift after cascaded FOPAs
Θ_s	Signal phase shift after cascaded FOPAs

κ_{fc}	Phase mismatch parameter of the frequency conversion process
κ_{pc}	Phase mismatch parameter of the phase conjugation process
κ_{sp}	Phase mismatch parameter of the single-pump process
λ	Wavelength
$\lambda_{ASE,peak}$	ASE peak wavelength
λ_c	Cut-off wavelength
λ_i	Converted signal (idler) wavelength
λ_p	Pump wavelength
λ_s	Input signal wavelength
λ_{zd}	Zero-dispersion wavelength
μ_0	Vacuum permeability
ξ	Sum phase
ρ	Charge density
ρ_c	Density of states in the conduction band
ρ_{AQN}	Power spectral density of additive white Gaussian noise
ρ_{ASE}	Power spectral density of the SOA amplified spontaneous emission noise
ρ_{AWG}	Power spectral density of additive white Gaussian noise
ρ_{QN}	Power spectral density of additive white Gaussian noise
ρ_{SE}	Power spectral density of the SOA spontaneous emission noise
σ	Conductivity

σ^2	Parameter of the log-normal distribution
σ_n^2	AWG noise variance
$\sigma_{amp,p}^2, \sigma_{amp,s}^2$	Pump and signal input amplitude variance
σ_{nan}^2	Nonlinear amplitude noise variance
σ_{nl}^2	Nonlinear noise variance
σ_{npn}^2	Nonlinear phase noise variance
$\sigma_{ph,p}^2, \sigma_{ph,s}^2$	Pump and signal phase variance
σ_{xpm}^2	Nonlinear noise variance due to XPM
$\sigma_{i,amp}, \sigma_{s,amp}$	Signal/idler amplitude standard deviation due to signal-induced noise
$\sigma_{i,ph}, \sigma_{s,ph}$	Signal/idler phase standard deviation due to signal-induced noise
τ	Arbitrary time interval
τ_{CH}	Time constant for carrier heating, free-carrier absorption and two-photon absorption
τ_{SHB}	Time constant for spectral hole burning
τ_s	Carrier lifetime
Φ_l	Total phase change in the l-th segment
ϕ_{CDP}	Phase change due to carrier density pulsations
ϕ_{CH}	Phase change due to carrier heating

ϕ_{FCA}	Phase change due to free carrier absorption
ϕ_k	Phase of the k-th symbol
$\tilde{\phi}_k$	Distorted phase of the k-th symbol
ϕ_{LO}	Local oscillator phase
ϕ_l	Laser phase noise
ϕ_{lpn}	Phase distortion due to laser phase noise
ϕ_p	Pump phase
ϕ_{ppm}	Phase distortion due to pump phase modulation
$\phi_{ppm,cd}$	Phase distortion due to pump phase modulation after carrier phase estimation
$\phi_{ppm,n}$	Contribution to the phase distortion due to the pump phase modulation after the n-th cascaded wavelength conversion
ϕ_{rpm}	Residual phase modulation
ϕ_{SHB}	Phase change due to spectral hole burning
ϕ_{sin}	Sinusoidal pump phase modulation
ϕ_{spm}	Phase distortion due to self-phase modulation
ϕ_{TPA}	Phase change due to two photon absorption
ϕ_{xpm}	Phase distortion due to cross-phase modulation
χ	Susceptibility
$\tilde{\chi}$	Frequency domain susceptibility

$\chi_{xx}^{(1)}$	Tensor element of the first order susceptibility
$\chi_{xxxx}^{(3)}$	Tensor element of the third order susceptibility
$\chi^{(n)}$	n-th order susceptibility
$\chi^{(r)}$	Weak field complex susceptibility
χ_p	Susceptibility due to pumping
χ_0	Susceptibility in absence of pumping
ψ_k	Recovered phase of the k-th symbol
Ω_B	Brillouin frequency shift
Ω_l	Difference optical angular frequency
Ω_R	Raman frequency shift
ω	Angular frequency
ω_a	Symmetry angular frequency
ω_c	Gain peak correction
ω_g	SOA band gap angular frequency
ω_i	Converted signal (idler) angular frequency
ω_p	Pump signal angular frequency
$\omega_{p,2}$	Gain peak frequency of quadratic estimation
$\omega_{p,3}$	Gain peak frequency of quadratic estimation
ω_s	Input signal angular frequency
ω_z	Begin of zero-gain region
ω_{zd}	Zero-dispersion angular frequency
$\omega_{z,0}$	Constant part of zero-gain region

ω_0	Optical reference angular frequency

Abbreviations

A/D	Analog-to-digital
AOWC	All-optical wavelength converter
ASE	Amplified spontaneous emission
AWG	Arrayed waveguide grating
AWG noise	Additive white Gaussian noise
BER	Bit error ratio
CB	Conduction band
CDP	Carrier density pulsation
CH	Carrier heating
CPE	Carrier phase estimation
CW	Continuous wave
DBPSK	Differential binary phase shift keying
DFG	Difference frequency generation
DFWM	Degenerate four-wave mixing
DPSK	Differential phase shift keying
DQPSK	Differential quadrature phase shift keying
EDE	Electronic distortion equalization
EDFA	Erbium doped fiber amplifier
FC	Frequency conversion
FCA	Free-carrier absorption
FOPA	Fiber optical parametric amplifier
FWM	Four-wave mixing
HNLF	Highly nonlinear fiber
InP	Indium phosphide
LO	Local oscillator
LP	Linearly polarized mode
LPF	Low-pass filter

NDFWM	Non-degenerate four-wave mixing
NF	Noise figure
OPC	Optical phase conjugation
OSNR	Optical signal-to-noise ratio
PC	Phase conjugation
PDF	Probability distribution function
PM	Phase modulator
PMD	Polarization-mode dispersion
PPLN	Periodically-poled lithium niobate
PSK	Phase shift keying
SBS	Stimulated Brillouin scattering
SEM	Symbol error rate
SGM	Self-gain modulation
SHB	Spectral-hole burning
SNR	Signal-to-noise ratio
SOA	Semiconductor optical amplifier
SOP	State of polarization
SP	Single pump
SPM	Self-phase modulation
SPR	Signal-to-pump ratio
SRS	Stimulated Raman scattering
SSMF	Standard single mode fiber
TE	Transverse electric
TM	Transverse magnetic
TPA	Two-photon absorption
QAM	Quadrature amplitude modulation
VB	Valence band
WDM	Wavelength division multiplex
XGM	Cross-gain modulation
XPM	Cross-phase modulation

Bibliography

[1] A. Ellis, J. Zhao, and D. Cotter, "Approaching the nonlinear shannon limit," *J. Lightw. Technol.*, vol. 28, no. 4, pp. 423–433, 2010.

[2] R. S. Tucker, "Green optical communications - part II: Energy limitations in networks." To be published in *J. Select. Topics of Quantum Electron.*, preprint available online with digital object identifier 10.1109/JSTQE.2010.2051217, 2010.

[3] J. Baliga, R. Ayre, K. Hinton, W. Sorin, and R. Tucker, "Energy consumption in optical IP networks," *J. Lightw. Technol.*, vol. 27, no. 13, pp. 2391–2403, 2009.

[4] S. Peng, K. Hinton, J. Baliga, R. Tucker, Z. Li, and A. Xu, "Burst switching for energy efficiency in optical networks," in *Proc. Optical Fiber Communications Conference (OFC)*, paper OWY5, 2010.

[5] I. Kaminov, T. Li, and A. Willner, eds., *Optical Fiber Telecommunications V-B*. Academic Press, 5th edition, San Diego, CA, 2008.

[6] M. Seimetz, *High-Order Modulation for Optical Fiber Transmission*. Springer-Verlag, Berlin, Germany, 2009.

[7] P. J. Roberts, F. Couny, H. Sabert, B. J. Mangan, D. P. Williams, L. Farr, M. W. Mason, A. Tomlinson, T. A. Birks,

J. C. Knight, and P. S. J. Russell, "Ultimate low loss of hollow-core photonic crystal fibres," *Opt. Express*, vol. 13, no. 1, p. 236, 2005.

[8] J. Boggio, S. Moro, B.-P. Kuo, N. Alic, B. Stossel, and S. Radic, "Tunable parametric all-fiber short-wavelength IR transmitter," *J. Lightw. Technol.*, vol. 28, no. 4, pp. 443–447, 2010.

[9] X. Liu, R. Osgood, Y. Vlasov, and W. Green, "Mid-infrared optical parametric amplifier using silicon nanophotonic waveguides," *Nature photonics*, vol. 4, pp. 557–560, 2010.

[10] P. L. Voss, R. Tang, and P. Kumar, "Measurement of the photon statistics and the noise figure of a fiber-optic parametric amplifier," *Opt. Lett.*, vol. 28, no. 7, pp. pp. 549 – 551, 2003.

[11] R. Tang, P. L. Voss, J. Lasri, P. Devgan, , and P. Kumar, "Noise-figure limit of fiber-optical parametric amplifiers and wavelength converters: experimental investigation," *Opt. Lett.*, vol. 29, no. 20, pp. 2372–2374, 2004.

[12] J. Mu and C. Savage, "Parametric amplifiers in phase-noise-limited optical communications," *J. Opt. Soc. Am. B*, vol. 9, no. 1, pp. 65–70, 1992.

[13] Z. Tong, C. McKinstrie, C. Lundström, M. Karlsson, and P. Andrekson, "Noise performance of optical fiber transmission links that use non-degenerate cascaded phase-sensitive amplifiers," *Opt. Express*, vol. 18, no. 15, pp. 15426–15439, 2010.

[14] Z. Tong, A. Bogris, C. Lundström, C. J. McKinstrie, M. Vasilyev, M. Karlsson, and P. Andrekson, "Zhi tong,1,* adonis bogris,2 carl lundström,1 c. j. mckinstrie,3 michael

vasilyev,4 magnus karlsson,1 and peter a. andrekson1," *Opt. Express*, vol. 18, no. 14, pp. 14820–14835, 2011.

[15] S. Jansen, D. van den Borne, P. Krummrich, S. Spälter, G.-D. Khoe, and H. de Waardt, "Long-haul DWDM transmission systems employing optical phase conjugation," *J. Select. Topics of Quantum Electron.*, vol. 12, no. 4, pp. 505–520, 2006.

[16] B. Kuo, E. Myslivets, A. Wiberg, S. Zlatanovic, C.-S. Brès, S. Moro, F. Gholami, A. Peric, N. Alic, and S. Radic, "Transmission of 640-Gb/s RZ-OOK channel over 100-km SSMF by wavelength-transparent conjugation," *J. Lightw. Technol.*, vol. 29, no. 4, pp. 516–523, 2011.

[17] P. Kaewplung and K. Kikuchi, "Simultaneous cancellation of fiber loss, dispersion, and Kerr effect in ultralong-haul optical fiber transmission by midway optical phase conjugation incorporated with distributed Raman amplification," *J. Lightw.Technol.*, vol. 25, no. 10, pp. 3035–3050, 2007.

[18] Y. Geng, C. Peucheret, and P. Jeppesen, "Amplitude equalization of 40 Gb/s RZ-DPSK signals using saturation of four-wave mixing in a highly nonlinear fiber," in *Proc. IEEE Lasers and Electro-Optics Society Annual Meeting (LEOS)*, paper MP5, 2006.

[19] K. Croussore and G. Li, "Phase and amplitude regeneration of differential phase-shift keyed signals using phase-sensitive amplification," *J. Sel. Topics. Quantum Electron.*, vol. 14, no. 3, pp. 648–658, 2008.

[20] F. Parmigiani, R. Slavic, J. Kakande, C. Lundström, M. Sjödin, P. Andrekson, R. Weerasuriya, S. Sygletos,

A. Ellis, L. Grüner-Nielsen, D. Jakobsen, S. Herstrom, R. Phelan, J. O'Gorman, A. Bogris, D. Syvridis, S. Dasgupta, P. Petropoulos, and D. Richardson, "All-optical phase regeneration of 40 Gbit/s DPSK signals in a blackbox phase sensitive amplifier," in *Proc. Optical Fiber Communications Conference (OFC)*, paper PDPC3, 2010.

[21] R. Slavik, F. Parmigiani, J. Kakande, C. Lundström, M. Sjödin, P. Andrekson, R. Weerasuriya, S. Sygletos, A. Ellis, L. Grüner-Nielsen, D. Jakobsen, S. Herstrom, R. Phelan, J. O'Gorman, A. Bogris, D. Syvridis, S. Dasgupta, P. Petropoulos, and D. Richardson, "All-optical phase and amplitude regenerator for next-generation telecommunications systems," *Nature photonics*, vol. 4, pp. 690–695, 2010.

[22] A. Fragkos, A. Bogris, and D. Syvridis, "All-optical regeneration based on phase-sensitive nondegenerate four-wave mixing in optical fibers," *IEEE Photon. Technol. Lett.*, vol. 22, no. 24, pp. 1826–1828, 2010.

[23] J. Hansryd and P. Andrekson, "Broad-band continuous-wave-pumped fiber optical parametric amplifier with 49-dB gain and wavelength-conversion efficiency," *IEEE Photon. Technol. Lett.*, vol. 13, no. 3, pp. 194–196, 2001.

[24] T. Torounidis, P. Andrekson, and B.-E. Olsson, "Fiber-optical parametric amplifier with 70-dB gain," *IEEE Photon. Technol. Lett.*, vol. 18, no. 10, p. 1194, 2006.

[25] S. Radic, C. McKinstrie, R. Jopson, J. Centanni, Q. Lin, and G. Agrawal, "Record performance of parametric amplifier constructed with highly nonlinear fibre," *Electron. Lett.*, vol. 39, no. 11, pp. 838–839, 2003.

[26] J. Chavez Boggio, J. D. Marconi, and H. Fragnito, "Double-pumped fiber optical parametric amplifier with flat gain over 47-nm bandwidth using a conventional dispersion-shifted fiber," *IEEE Photon. Technol. Lett.*, vol. 17, no. 9, pp. 1842–1844, 2005.

[27] J. D. Marconi, J. M. Chavez Boggio, H. L. Fragnito, and S. Bickham, "Nearly 100 nm bandwidth of flat gain with a double-pumped fiber optic parametric amplifier," in *Optical Fiber Communications Conference (OFC)*, paper OWB1, 2007.

[28] R. Jiang, N. Alic, C. McKinstrie, and S. Radic, "Two pump parametric amplifier with 40dB of equalized continuous gain over 50nm," in *Optical Fiber Communication Conference (OFC)*, paper OWB2, 2007.

[29] R. Tucker, "The role of optics and electronics in high-capacity routers," *J. Lightw. Technol.*, vol. 24, no. 12, pp. 4655–4673, 2006.

[30] S. J. B. Yoo, "Optical packet and burst switching technologies for the future photonic internet," *J. Lightw. Technol.*, vol. 24, no. 12, pp. 4468–4492, 2006.

[31] E. Myslivets, N. Alic, S. Moro, B. Kuo, R. Jopson, C. McKinstrie, M. Karlsson, and S. Radic, "1.56-μs continuously tunable parametric delay line for a 40-Gb/s signal," *Opt. Express*, vol. 17, no. 14, pp. 11958–11964, 2009.

[32] C. McKinstrie, S. Radic, R. Jopson, and A. Chraplyvy, "Quantum noise limits on optical monitoring with parametric devices," *Opt. Communications*, vol. 259, no. 1, pp. 309–320, 2006.

[33] L. Ceipidor, A. Bosco, and E. Fazio, "Logic functions, devices, and circuits based on parametric nonlinear processes," *J. Lightw. Technol.*, vol. 26, no. 3, pp. 373–378, 2008.

[34] S. Yoo, "Wavelength conversion technologies for WDM network applications," *J. Lightw. Technol.*, vol. 14, no. 6, pp. 955–966, 1996.

[35] X. Yi, R. Yu, J. Kurumida, and S. Yoo, "Modulation-format-independent wavelength conversion," in *Optical Fiber Communication Conference (OFC)*, paper PDPC9, 2009.

[36] X. Yi, R. Yu, J. Kurumida, and S. Yoo, "A theoretical and experimental study on modulation-format-independent wavelength conversion," *J. Lightw. Technol.*, vol. 28, no. 4, pp. 587–595, 2010.

[37] R. Elschner, C. Bunge, and K. Petermann, "System impact of cascaded all-optical wavelength conversion of D(Q)PSK signals in transparent optical networks," *Journal of Networks*, vol. 5, no. 2, pp. 219–224, 2010.

[38] T. Umeki, O. Tadanaga, and M. Asobe, "Highly efficient wavelength converter using direct-bonded PPZnLN ridge waveguide," *J. Quantum Electron.*, vol. 46, no. 8, pp. 1206–1213, 2010.

[39] T. Umeki, O. Tadanaga, A. Takada, and M. Asobe, "Phase sensitive degenerate parametric amplification using directly-bonded PPLN ridge waveguides," *Opt. Express*, vol. 19, no. 7, pp. 6326–6332, 2011.

[40] M. Takahashi, R. Sugizaki, J. Hiroishi, M. Tadakuma, Y. Taniguchi, and T. Yagi, "Low-loss and low-dispersion-

slope highly nonlinear fibers," *J. Lightw. Technol.*, vol. 23, no. 11, pp. 3615–3624, 2005.

[41] T. Okuno, T. Nakanishi, M. Hirano, and M. Onishi, "Practical considerations for the application of highly nonlinear fibers," in *Optical Fiber Communications Conference (OFC)*, paper OTuJ1, 2006.

[42] R. Elschner, C.-A. Bunge, B. Hüttl, A. Gual i Coca, C. Schmidt-Langhorst, R. Ludwig, C. Schubert, and K. Petermann, "Impact of pump-phase modulation on FWM-based wavelength conversion of D(Q)PSK signals," *J. Sel. Topics Quantum Electron.*, vol. 14, no. 3, pp. 666–673, 2008.

[43] R. Elschner, C.-A. Bunge, and K. Petermann, "Co- and counterphasing tolerances for dual-pump parametric λ-conversion of D(Q)PSK signals," *IEEE Photon. Technol. Lett.*, vol. 21, no. 11, pp. 706–708, 2009.

[44] R. Elschner, T. Richter, and K. Petermann, "Impact of pump-phase modulation on fibre-based parametric wavelength conversion of coherently detected PSK signals," in *Proc. Photonics in Switching (PiS)*, paper FrI1-5, 2009.

[45] R. Elschner, T. Richter, L. Molle, K. Petermann, and C. Schubert, "Single-pump FWM-wavelength conversion in HNLF using coherent receiver-based electronic compensation," in *Proc. European Conference on Optical Communications (ECOC)*, paper P3.17, 2010.

[46] R. Elschner and K. Petermann, "Impact of pump-induced nonlinear phase noise on parametric amplification and wavelength conversion of phase-modulated signals," in

Proc. European Conference on Optical Communications (ECOC), paper 3.3.4, 2009.

[47] R. Elschner and K. Petermann, "BER performance of 16-QAM signals amplified by dual-pump fiber optical parametric amplifiers," in *Proc. Optical Fiber Communications Conference (OFC)*, paper OThA4, 2010.

[48] R. Elschner and K. Petermann, "Pump-induced nonlinear phase noise in wavelength converters based on four-wave mixing in SOAs," in *Proc. IEEE Photonics Society Annual Meeting (former LEOS)*, paper ThU4, 2009.

[49] R. Boyd, *Nonlinear Optics*. Academic Press, 3rd edition, San Diego, CA, 2008.

[50] G. Agrawal, *Nonlinear Fiber Optics*. Academic Press, 4th edition, San Diego, CA, 2006.

[51] G. Agrawal and N. Dutta, *Long-Wavelength Semiconductor Lasers*. Van Nostrand Reinhold Company, New York, NY, 1986.

[52] M.-J. Li, S. Li, and D. A. Nolan, "Nonlinear fiber for signal processing using optical Kerr effects," *J. Lightw. Technol.*, vol. 23, no. 11, pp. 3606–3614, 2005.

[53] J. Gowar, *Optical Communication Systems*. Prentice Hall, Upper Saddle River, NJ, 1984.

[54] J. M. Senior, *Optical Fiber Communications*. Prentice Hall, Upper Saddle River, NJ, 1985.

[55] A. W. Snyder and J. D. Love, *Optical Waveguide Theory*. Chapman and Hall, London, Great Britain, 1983.

[56] T. Okuno, M. Onishi, M. Kashiwada, S. Ishikawa, and M. Nishimura, "Silica-based functional fibers with enhanced nonlinearity and their applications," *IEEE J. Sel. Topics Quantum Electron.*, vol. 5, no. 5, pp. 1385–1391, 1999.

[57] www.photonics.umd.edu/software/ssprop.

[58] O. V. Sinkin, R. Holzlöhner, J. Zweck, and C. R. Menyuk, "Optimization of the split-step Fourier method in modeling optical-fiber communications systems," *J. Lightw. Technol.*, vol. 21, no. 1, pp. 61–68, 2003.

[59] A. Kobyakov, M. Mehendale, M. Vasilyev, S. Tsuda, and A. Evans, "Stimulated Brillouin scattering in Raman-pumped fibers: A theoretical approach," *J. Lightw. Technol.*, vol. 20, no. 8, pp. 1635–1643, 2002.

[60] R. G. Smith, "Optical power handling capacity of low loss optical fibers as determined by stimulated Raman and Brillouin scattering," *Appl. Opt.*, vol. 11, no. 11, pp. 2489–2494, 1972.

[61] A. Mussot, E. Lantz, A. Durécu-Legrand, C. Simonneau, D. Bayart, T. Sylvestre, and H. Maillotte, "Zero-dispersion wavelength mapping in short single-mode optical fibers using parametric amplification," *IEEE Photon. Technol. Lett.*, vol. 18, no. 1, pp. 22–24, 2006.

[62] M. Hirano, T. Nakanishi, and T. Sasaki, "FWM-based flexible wavelength conversion in whole C-band using realistic HNLF having dispersion slope," in *Proc. Optical Fiber Communications Conference (OFC)*, paper OTuA2, 2010.

[63] C. H. Henry, "Theory of the linewidth of semiconductor lasers," *J. Quantum Electron.*, vol. 18, no. 2, pp. 259–264, 1982.

[64] A. Uskov, J. Mork, and J. Mark, "Wave mixing in semiconductor laser amplifiers due to carrier heating and spectral-hole burning," *J. Quantum Electron.*, vol. 30, no. 8, pp. 1769–1781, 1994.

[65] K. Hall, G. Lenz, and E. Ippen, "Femtosecond time domain measurements of group velocity dispersion in diode lasers at 1.5 μm," *J. Lightw. Technol.*, vol. 10, no. 5, pp. 616–619, 1992.

[66] P. Runge, R. Elschner, and K. Petermann, "Chromatic dispersion in InGaAsP semiconductor optical amplifiers," *J. Quantum Electron.*, vol. 46, no. 5, pp. 644–649, 2010.

[67] C. Holtmann, P. Besse, T. Brenner, and H. Melchior, "Polarization independent bulk active region semiconductor optical amplifiers for 1.3 μm wavelengths," *IEEE Photon. Technol. Lett.*, vol. 8, no. 3, pp. 343–345, 1996.

[68] M. Connelly, *Semiconductor Optical Amplifiers*. Kluwer Academic, Amsterdam, Netherlands, 2000.

[69] J. Mark and J. Mork, "Subpicosecond gain dynamics in InGaAsP optical amplifiers: Experiment and theory," *Appl. Phys. Lett.*, vol. 61, no. 19, pp. 2281–2283, 1992.

[70] K. Hall, G. Lenz, A. Darwish, and E. Ippen, "Subpicosecond gain and index nonlinearities in InGaAsP diode lasers," *Opt. Communications*, vol. 111, no. 5-6, pp. 589–612, 1994.

[71] A. Mecozzi and J. Mork, "Saturation effects in nondegenerate four-wave mixing between short optical pulses in semi-

conductor laser amplifiers," *J. Sel. Topics Quantum Electron.*, vol. 3, no. 5, pp. 1190–1207, 1997.

[72] P. Runge, R. Elschner, and K. Petermann, "Time-domain modelling of ultralong semiconductor optical amplifiers," *J. Quantum. Electron*, vol. 46, no. 5, pp. 484–491, 2010.

[73] J. Leuthold, M. Mayer, J. Eckner, G. Guekos, H. Melchior, and C. Zellweger, "Material gain of bulk 1.55 µm InGaAsP/InP semiconductor optical amplifiers approximated by a polynomial model," *J. Applied Physics*, vol. 87, no. 1, pp. 618–620, 2000.

[74] G. Toptchiyski, S. Kindt, K. Petermann, E. Hilliger, S. Diez, and H. Weber, "Time-domain modeling of semiconductor optical amplifiers for OTDM applications," *J. Lightw. Technol.*, vol. 17, no. 12, pp. 2577–2583, 1999.

[75] A. M. Melo and K. Petermann, "On the amplified spontaneous emission noise modeling of semiconductor optical amplifiers," *Opt. Communications*, vol. 281, no. 18, pp. 4598–4605, 2008.

[76] J. Mork and J. Mark, "Time-resolved spectroscopy of semiconductor laser devices: Experiments and modeling," *Proc. SPIE*, vol. 2399, pp. 146–159, 1995.

[77] R.-J. Essiambre, G. Kramer, P. Winzer, G. Foschini, and B. Goebel, "Capacity limits of optical fiber networks," *J. Lightw. Technol.*, vol. 28, no. 4, pp. 662–701, 2010.

[78] E. Ip, A. P. T. Lau, D. Barros, and J. Kahn, "Coherent detection in optical fiber systems," *Opt. Express*, vol. 16, no. 2, pp. 753–791, 2008.

[79] K. Ho and H.-W. Cuei, "Generation of arbitrary quadrature signals using one dual-drive modulator," *J. Lightw. Technol.*, vol. 22, no. 3, pp. 764–770, 2005.

[80] K. Ho, *Phase Modulated Optical Communication Systems*. Springer-Verlag, Berlin, Germany, 2005.

[81] N. Blachman, "The effect of phase error on DPSK error probability," *IEEE Transactions Commun.*, vol. 29, no. 3, pp. 364–365, 1981.

[82] K. Ho, "Impact of nonlinear phase noise to DPSK signals:A comparison of different models," *IEEE Photon. Technol. Lett.*, vol. 16, no. 5, pp. 1403–1405, 2004.

[83] J. Hansryd, P. Andrekson, M. Westlund, J. Li, and P.-O. Hedekvist, "Fiber-based optical parametric amplifiers and their applications," *IEEE J. Sel. Topics Quantum Electron.*, vol. 8, no. 3, pp. 506–520, 2002.

[84] P. Kylemark, P. Hedekvist, H. Sunnerud, M. Karlsson, and P. Andrekson, "Noise characteristics of fiber optical parametric amplifiers," *J. Lightw. Technol.*, vol. 22, no. 2, pp. 409–416, 2004.

[85] C. McKinstrie, S. Radic, and A. Chraplyvy, "Parametric amplifiers driven by two pump waves," *IEEE J. Sel. Topics Quantum Electron.*, vol. 8, no. 3, pp. 538–547, 2002.

[86] M. Takahashi, S. Takasaka, R. Sugizaki, and T. Yagi, "Arbitrary wavelength conversion in entire CL-band based on pump-wavelength-tunable FWM in a HNLF," in *Proc. Optical Fiber Communications Conference (OFC)*, paper OWP4, 2010.

[87] E. Myslivets, C. Lundström, S. Moro, A. Wiberg, C.-S. Bres, N. Alic, P. Andrekson, and S. Radic, "Dispersion fluctuation

equalization nonlinear fibers by spatially controlled tension," in *Proc. Optical Fiber Communications Conference (OFC)*, paper OTuA5, 2010.

[88] H. Haus, "The noise figure of optical amplifiers," *IEEE Photon. Technol. Lett.*, vol. 10, no. 11, p. 1602, 1998.

[89] A. Bogris, D. Syvridis, P. Kylemark, and P. Andrekson, "Noise characteristics of dual-pump fiber-optic parametric amplifiers," *J. Lightw. Technol.*, vol. 23, no. 9, pp. 2788–2795, 2005.

[90] C. McKinstrie, M. Yu, M. G. Raymer, and S. Radic, "Quantum noise properties of parametric processes," *Opt. Express*, vol. 13, no. 13, pp. 4986–5012, 2005.

[91] P. L. Voss and P. Kumar, "Raman-noise-induced noise-figure limit for $\chi^{(3)}$ parametric amplifiers," *Opt. Lett.*, vol. 29, no. 5, pp. 445–447, 2004.

[92] C. McKinstrie and S. Radic, "Phase-sensitive amplification in a fiber," *Opt. Express*, vol. 12, no. 20, pp. 4973–4979, 2004.

[93] P. Kylemark, M. Karlsson, T. Torounidis, and P. Andrekson, "Noise statistics in fiber optical parametric amplifiers," *J. Lightw. Technol.*, vol. 25, no. 2, pp. 612–620, 2007.

[94] K. Wong, M. Marhic, and L. Kazovsky, "Phase-conjugate pump dithering for high-quality idler generation in a fiber optical parametric amplifier," *IEEE Photon. Technol. Lett.*, vol. 15, no. 1, pp. 33–35, 2003.

[95] K. Shiraki, M. Ohashi, and M. Tateda, "Suppression of stimulated Brillouin scattering in a fibre by changing the

core radius," *Electron. Lett.*, vol. 31, no. 8, pp. 668–669, 1995.

[96] K. Shiraki, M. Ohashi, and M. Tateda, "Performance of strain-free stimulated Brillouin scattering suppression fiber," *J. Lightw. Technol.*, vol. 14, pp. 549–554, 1996.

[97] K. Tsujikawa, K. Nakajima, Y. Miyajima, and M. Ohashi, "New SBS suppression fiber with uniform chromatic dispersion to enhance four-wave mixing," *IEEE Photon. Technol. Lett.*, vol. 10, no. 8, pp. 1139–1141, 1998.

[98] N. Yoshizawa, T. Horiguchi, and T. Kurashima, "Proposal for stimulated Brillouin scattering suppression by fibre cabling," *Electron. Lett.*, vol. 27, no. 12, pp. 1100–1101, 1991.

[99] T. Horiguchi, T. Kurashima, and M. Tateda, "Tensile strain dependence of Brillouin frequency shift in silica optical fibers," *IEEE Photon. Technol. Lett.*, vol. 1, no. 5, pp. 107–108, 1989.

[100] J. Hansryd, F. Dross, M. Westlund, P. Andrekson, and S. Knudsen, "Increase of the SBS threshold in a short highly nonlinear fiber by applying a temperature distribution," *J. Lightw. Technol.*, vol. 19, no. 11, pp. 1691–1697, 2001.

[101] Y. Imai and N. Shimada, "Dependence of stimulated Brillouin scattering on temperature distribution in polarization-maintaining fibers," *IEEE Photon. Technol. Lett.*, vol. 5, no. 11, pp. 1335–1337, 1993.

[102] Y. Aoki, K. Tajima, and I. Mito, "Input power limits of single-mode optical fibers due to stimulated Brillouin scattering in optical communication systems," *J. Lightw. Technol.*, vol. 6, no. 5, pp. 710–719, 1988.

[103] S. K. Korotky, P. B. Hansen, L. Eskildsen, and J. J. Veselka, "Efficient phase modulation scheme for suppressing stimulated Brillouin scattering," in *Proc. Integrated Optics and Optical Fiber Communications (IOOC)*, paper WD2, 1995.

[104] J. Yu and M.-F. Huang, "Wavelength conversion based on copolarized pumps generated by optical carrier suppression," *IEEE Photon. Technol. Lett.*, vol. 21, no. 6, pp. 392–394, 2009.

[105] S. Watanabe, T. Kato, R. Okabe, R. Elschner, R. Ludwig, and C. Schubert, "All-optical data frequency multiplexing on single-wavelength carrier light by sequentially provided cross-phase modulation in fiber." To be published in *J. Select. Topics of Quantum Electron.*, preprint available online with digital object identifier 10.1109/JSTQE.2011.2111358, 2011.

[106] B. Huettl, A. Gual i Coca, R. Elschner, C.-A. Bunge, K. Petermann, C. Schmidt-Langhorst, R. Ludwig, and C. Schubert, "Optimization of SBS-suppression for 320 Gbit/s DQPSK all-optical wavelength conversion," in *Proc. European Conference on Optical Communication (ECOC)*, paper 4.5.5, 2007.

[107] C.-A. Bunge, R. Elschner, P. Runge, and K. Petermann, "All-optical wavelength conversion of D(Q)PSK signals in transparent optical networks," in *Proc. International Conference on Transparent Optical Networking (ICTON)*, 2008.

[108] M.-C. Ho, M. Marhic, K. Wong, and L. Kazovsky, "Narrow-linewidth idler generation in fiber four-wave mixing and parametric amplification by dithering two pumps in opposition of phase," *J. Lightw. Technol.*, vol. 20, no. 3, pp. 469–476, 2002.

[109] T. Richter, R. Elschner, K. Petermann, and C. Schubert, "Tolerances of counterphased pump-phase modulation in a fibre- based dual-pump wavelength converter for 86 Gb/s RZ-DQPSK," in *Proc. Photonics in Switching (PiS)*, paper FrI1-3, 2009.

[110] N. Alic, R. M. Jopson, J. Ren, E. Myslivets, R. Jiang, A. H. Gnauck, and S. Radic, "Impairments in deeply-saturated optical parametric amplifiers for amplitude- and phase-modulated signals," *Opt. Express*, vol. 15, no. 14, pp. 8997–9008, 2007.

[111] T. Richter, R. Elschner, K. Petermann, and C. Schubert, "Fibre-based parametric wavelength conversion of 86 Gb/s RZ-DQPSK signals with 15 dB gain using a dual-pump configuration," in *Proc. European Conference on Optical Communications (ECOC)*, paper 3.3.2, 2009.

[112] K. Torii and S. Yamashita, "Efficiency improvement of optical fiber wavelength converter without spectral spread using synchronous phase/frequency modulations," *J. Lightw. Technol.*, vol. 21, no. 4, pp. 1039–1045, 2003.

[113] S. Yamashita and M. Tani, "Cancellation of spectral spread in SBS-suppressed fiber wavelength converters using a single phase modulator," *IEEE Photon. Technol. Lett.*, vol. 16, no. 9, pp. 2096–2098, 2004.

[114] O. Yilmaz, J. Wang, S. Khaleghi, X. Wang, S. Nuccio, X. Wu, and A. Willner, "Preconversion phase modulation of input differential phase-shift-keying signals for wavelength conversion and multicasting applications using phase-modulated pumps," *Opt. Lett.*, vol. 36, no. 5, pp. 731–733, 2011.

[115] T. Richter, R. Elschner, L. Molle, K. Petermann, and C. Schubert, "Coherent receiver-based compensation of phase distortions induced by single-pump HNLF-based FWM wavelength converters," in *Proc. Photonics in Switching (PiS)*, paper PWB2, 2010.

[116] M. Abe and J. I. Smith, "Design criteria for the quadratically interpolated FFT method (I): Bias due to interpolation," in *Stanford Univ., Stanford, CA, Tech. Rep. STAN-M-114*, 2004.

[117] A. Mussot, A. D. Legrand, E. Lantz, C. Simonneau, D. Bayart, H. Maillotte, and T. Sylvestre, "Impact of pump phase modulation on the gain of fiber optical parametric amplifier," *IEEE Photon. Technol. Lett.*, vol. 16, no. 5, pp. 1289–1291, 2004.

[118] A. Durecu-Legrand, A. Mussot, C. Simonneau, D. Bayart, T. Sylvestre, E. Lantz, and H. Maillotte, "Impact of pump phase modulation on system performance of fibre-optical parametric amplifiers," *Electron. Lett.*, vol. 41, no. 6, pp. 350–352, 2005.

[119] P. Kylemark, *Noise and Saturation Properties of Fiber Optical Parametric Amplifiers*. PhD thesis, Chalmers University of Technology, Göteborg, Sweden, 2006.

[120] F. Yaman, Q. Lin, S. Radic, and G. Agrawal, "Impact of pump-phase modulation on dual-pump fiber-optic parametric amplifiers and wavelength converters," *IEEE Photon. Technol. Lett.*, vol. 17, no. 10, pp. 2053–2055, 2005.

[121] P. Kylemark, H. Sunnerud, M. Karlsson, and P. Andrekson, "Semi-analytic saturation theory of fiber optical paramet-

ric amplifiers," *J. Lightw. Technol.*, vol. 24, no. 9, pp. 3471–3479, 2006.

[122] P. Kylemark, J. Ren, Y. Myslivets, N. Alic, S. Radic, P. A. Andrekson, and M. Karlsson, "Impact of pump phase-modulation on the bit-error rate in fiber-optical parametric-amplifier-based systems," *IEEE Photon. Technol. Lett.*, vol. 19, no. 2-4, pp. 79–81, 2007.

[123] J. M. C. Boggio and H. L. Fragnito, "Simple four-wave-mixing-based method for measuring the ratio between the third- and fourth-order dispersion in optical fibers," *J. Opt. Soc. Am. B*, vol. 24, no. 9, pp. 2046–2054, 2007.

[124] P. Kylemark, M. Karlsson, and P. Andrekson, "Impact of phase modulation and filter characteristics on dual-pumped fiber-optical parametric amplification," *IEEE Photon. Technol. Lett.*, vol. 18, no. 2, pp. 439–441, 2006.

[125] M. Sköld, J. Yang, H. Sunnerud, M. Karlsson, S. Odaand, and P. Andrekson, "Constellation diagram analysis of DPSK signal regeneration in a saturated parametric amplifier," *Opt. Express*, vol. 16, no. 9, pp. 5974–5982, 2008.

[126] M. Matsumoto, "Phase noise generation in an amplitude limiter using saturation of a fiber-optic parametric amplifier," *Opt. Lett.*, vol. 33, no. 15, pp. 1638–1640, 2008.

[127] S. Moro, A. Danicic, N. Alic, B. Stossel, and S. Radic, "Noise-induced nonlinear frequency chirping in $\chi^{(3)}$ nonlinear media," *Opt. Express*, vol. 18, no. 22, pp. 23413–23419, 2010.

[128] S. Moro, A. Peric, N. Alic, B. Stossel, and S. Radic, "Phase noise in fiber-optic parametric amplifiers and converters

and its impact on sensing and communication systems," *Opt. Express*, vol. 18, no. 20, pp. 21449–21460, 2010.

[129] J. Proakis, *Digital communications*. Mcgraw-Hill, 4th edition, New York, NY, 2000.

[130] F. Yaman, Q. Lin, G. Agrawal, and S. Radic, "Pump-noise transfer in dual-pump fiber-optic parametric amplifiers: Walk-off effects," *Opt. Lett.*, vol. 30, no. 9, pp. 1048–1050, 2005.

[131] R. Elschner, T. Richter, M. Nölle, J. Hilt, and C. Schubert, "Parametric amplification of 28-GBd NRZ-16QAM signals," in *Proc. Optical Fiber Communications Conference (OFC)*, paper OThC2, 2011.

[132] H. Steffensen, J. Ott, K. Rottwitt, and C. McKinstrie, "Full and semi-analytic analyses of two-pump parametric amplification with pump depletion," *Opt. Express*, vol. 19, no. 7, pp. 6648–6656, 2011.

[133] F. Girardin, J. Eckner, G. Guekos, R. Dall'Ara, A. Mecozzi, A. D'Ottavi, F. Martelli, S. Scotti, and P. Spano, "Low-noise and very high-efficiency four-wave mixing in 1.5-μm-long semiconductor optical amplifiers," *IEEE Photon. Technol. Lett.*, vol. 9, no. 6, pp. 746–748, 1997.

[134] A. D'Ottavi, F. Girardin, L. Graziani, F. Martelli, P. Spano, A. Mecozzi, S. Scotti, R. Dall'Ara, J. Eckner, and G. Guekos, "Four-wave mixing in semiconductor optical amplifiers: A practical tool for wavelength conversion," *J. Sel. Topics Quantum Electron.*, vol. 3, no. 2, pp. 522–528, 1997.

[135] D. Geraghty, R. Lee, M. Verdiell, M. Ziari, A. Mathur, and K. Vahala, "Wavelength conversion for WDM communication systems using four-wave mixing in semiconductor op-

tical amplifiers," *J. Sel. Topics Quantum Electron.*, vol. 3, no. 5, pp. 1146–1155, 1997.

[136] S. Diez, C. Schmidt, R. Ludwig, H. Weber, K. Obermann, S. Kindt, I. Koltchanov, and K. Petermann, "Four-wave mixing in semiconductor optical amplifiers for frequency conversion and fast optical switching," *J. Sel. Topics Quantum Electron.*, vol. 3, no. 5, pp. 1131–1145, 1997.

[137] T. Morgan, J. Lacey, and R. Tucker, "Widely tunable four-wave mixing in semiconductor optical amplifiers with constant conversion efficiency," *IEEE Photon. Technol. Lett.*, vol. 10, no. 10, pp. 1401–1403, 1998.

[138] J. Lacey, M. Summerfield, and S. J. Madden, "Tunability of polarization-insensitive wavelength converters based on four-wave mixing in semiconductor optical amplifiers," *J. Lightw. Technol.*, vol. 16, no. 12, pp. 2419–2427, 1998.

[139] G. Contestabile, L. Banchi, M. Presi, and E. Ciaramella, "Investigation of transparency of FWM in SOA to advanced modulation formats involving intensity, phase, and polarization multiplexing," *J. Lightw. Technol.*, vol. 27, no. 19, pp. 4256–4261, 2009.

[140] L. Han, H. Hu, R. Ludwig, C. Schubert, and H. Zhang, "All-optical wavelength conversion of 80 Gb/s RZ-DQPSK using four-wave mixing in a semiconductor optical amplifier," in *Proc. IEEE Lasers and Electro-Optics Society Annual Meeting (LEOS)*, paper MP4, 2008.

[141] A. Mecozzi, A. D'Ottavi, F. Cara Romeo, P. Spano, R. Dall'Ara, G. Guekos, and J. Eckner, "High saturation behavior of the four-wave mixing signal in semiconductor

amplifiers," *Appl. Phys. Lett.*, vol. 66, no. 10, pp. 1184–1186, 1994.

[142] K. Obermann, I. Koltchanov, K. Petermann, S. Diez, R. Ludwig, and H. Weber, "Noise analysis of frequency converters utilizing semiconductor-laser amplifiers," *J. Quantum Electron.*, vol. 33, no. 1, pp. 81–88, 1997.

[143] H. Hu, L. Han, R. Ludwig, C. Schmidt-Langhorst, J. Yu, and C. Schubert, "107 Gb/s RZ-DQPSK signal transmission over 108 km SMF using optical phase conjugation in a SOA," in *Proc. Optical Fiber Communications Conference (OFC)*, paper OThF6, 2009.

[144] J. Leuthold, C. Koos, and W. Freude, "Nonlinear silicon photonics," *Nature photonics*, vol. 4, pp. 535–544, 2010.

[145] P. Juodawlkis, J. Plant, W. Loh, L. Missaggia, K. Jensen, and F. O'Donnell, "Packaged 1.5-μm quantum-well SOA with 0.8-W output power and 5.5-dB noise figure," *IEEE Photon. Technol. Lett.*, vol. 21, no. 17, pp. 1208–1210, 2009.

[146] W. Loh, J. Plant, J. Klamkin, J. Donnelly, F. O'Donnell, R. Ram, and P. Juodawlkis, "Noise figure of watt-class ultralow-confinement semiconductor optical amplifiers," *J. Quantum Electron.*, vol. 47, no. 1, p. 66, 2011.

[147] H. Hu, R. Nouroozi, R. Ludwig, B. Huettl, C. Schmidt-Langhorst, H. Suche, W. Sohler, and C. Schubert, "Simultaneous polarization-insensitive wavelength conversion of 80-Gb/s RZ-DQPSK signal and 40-Gb/s RZ-OOK signal in a Ti:PPLN waveguide," *J. Lightw. Technol.*, vol. 29, no. 8, pp. 1092–1097, 2011.

[148] H. Hu, R. Nouroozi, R. Ludwig, C. Schmidt-Langhorst, H. Suche, W. Sohler, and C. Schubert, "Polarization insen-

sitive 320 Gb/s in-line all-optical wavelength conversion in a 320-km transmission span." To be published in IEEE Photon. Technol. Lett., preprint available online with digital object identifier 10.1109/LPT.2011.2119473, 2011.

[149] T. Richter, R. Elschner, A. Gandhe, and C. Schubert, "Parametric amplification of 112 Gbit/s polarization multiplexed DQPSK signals in a fiber loop configuration," in *Proc. Optical Fiber Communications Conference (OFC)*, paper OThC4, 2011.

[150] N. El Dahdah, D. Govan, M. Jamshidifar, N. Doran, and M. Marhic, "1-Tb/s DWDM long-haul transmission employing a fiber optical parametric amplifier," *IEEE Photon. Technol. Lett.*, vol. 22, no. 15, pp. 1171–1173, 2010.

[151] X. Li and G. Li, "Electrical postcompensation of SOA impairments for fiber-optic transmission," *IEEE Photon. Technol. Lett.*, vol. 21, no. 9, pp. 581–583, 2009.

[152] X. Li and G. Li, "Joint fiber and soa impairment compensation using digital backward propagation," *IEEE Photon. J.*, vol. 2, no. 5, pp. 753–758, 2010.

[153] F. Vacondio, A. Ghazisaeidi, A. Bononi, and L. Rusch, "Low-complexity compensation of SOA nonlinearity for single-channel PSK and OOK," *J. Lightw. Technol.*, vol. 28, no. 3, pp. 277–286, 2010.

[154] P. Juodawlkis, W. Loh, F. O'Donnell, M. Brattain, and J. Plant, "High-power, ultralow-noise semiconductor external cavity lasers based on low-confinement optical waveguide gain media," *Proc. of SPIE*, vol. 7616, pp. 76160X–1 – 76160X–9, 2010.

i want morebooks!

Buy your books fast and straightforward online - at one of world's fastest growing online book stores! Environmentally sound due to Print-on-Demand technologies.

Buy your books online at
www.get-morebooks.com

Kaufen Sie Ihre Bücher schnell und unkompliziert online – auf einer der am schnellsten wachsenden Buchhandelsplattformen weltweit! Dank Print-On-Demand umwelt- und ressourcenschonend produziert.

Bücher schneller online kaufen
www.morebooks.de

VDM Verlagsservicegesellschaft mbH
Heinrich-Böcking-Str. 6-8
D - 66121 Saarbrücken

Telefon: +49 681 3720 174
Telefax: +49 681 3720 1749

info@vdm-vsg.de
www.vdm-vsg.de

Printed by Books on Demand GmbH, Norderstedt / Germany